I0069993

Building Competitive Firms
Incentives and Capabilities

Ijaz Nabi and Manjula Luthria
Editors

THE WORLD BANK
Washington, D.C.

© 2002 The International Bank for Reconstruction and Development/THE WORLD BANK
1818 H Street, NW
Washington, DC 20433 USA

Telephone   202-473-1000
Internet   www.worldbank.org
E-mail   feedback@worldbank.org

All rights reserved
First printing September 2002

The findings, interpretations, and conclusions expressed here are those of the authors and do not necessarily reflect the views of the Board of Executive Directors of the World Bank or the governments they represent.

The World Bank does not guarantee the accuracy of the data included in this work. The boundaries, colors, denominations, and other information shown on any map in this work do not imply any judgment on the part of the World Bank concerning the legal status of any territory or the endorsement or acceptance of such boundaries.

Rights and Permissions
The material in this work is copyrighted. Copying and/or transmitting portions or all of this work without permission may be a violation of applicable law. The World Bank encourages dissemination of its work and will normally grant permission promptly.

For permission to photocopy or reprint any part of this work, please send a request with complete information to the Copyright Clearance Center, Inc., 222 Rosewood Drive, Danvers, MA 01923, USA, telephone 978-750-8400, fax 978-750-4470, www.copyright.com.

All other queries on rights and licenses, including subsidiary rights, should be addressed to the Office of the Publisher, World Bank, at the above address, by fax 202-522-2422, or e-mail pubrights@worldbank.org.

Cover design: Seriph Design Group
Cover photo: Alex Mares-Manton/Getty Images

ISBN 0-8213-5154-0

Library of Congress Cataloging-in-Publication Data has been requested

POD by LSI

Contents

Figures

Boxes

Preface

In today's global economy firms face an increasingly competitive trade and production environment that requires continued vigilance to stay ahead of rivals. Competitiveness is therefore a major preoccupation of national, regional, and local governments worldwide as they seek to strengthen the investment climate in their jurisdictions. There is a growing recognition that competitiveness requires more than a liberal trade environment—that domestic policies and institutions, the so-called "behind-the-border" issues, are critical for countries that wish to reap the full benefits of trade liberalization.

While firm-level competitiveness is influenced by a host of macro- and micro-economic factors, the focus of this book is micro-level determinants. The first chapter presents a conceptual framework and a context for the key determinants of competitiveness, drawing on the recent East Asian experience. The eight following chapters provide in-depth discussion that helps deepen the understanding of micro-level determinants in policy and institutional settings. We hope that these discussions will contribute to the thinking about key issues in the operational restructuring of corporations, which remains a challenge as corporations struggle with post-crisis financial restructuring.

We would like to thank the Asia-Europe Meeting (ASEM) trust fund donors for funding the work on which this book has drawn. ASEM resources also facilitated the hosting of discussion workshops in Korea, the Philippines, Singapore, and Thailand, which were attended by senior policymakers in these countries. We are grateful to the World Bank's country and sectoral teams, without whose support the workshops and subsequent policy dialogue would not have been possible. We would also like to thank the authors for cheerfully meeting our many editorial demands. The encouragement of Homi Kharas, Sector Director of the Poverty Reduction and Economic Management Unit in the East Asia and Pacific Region of the Bank, is gratefully acknowledged.

1
Building Competitiveness: A Roadmap and Policy Guide

Ijaz Nabi and Manjula Luthria

During the 1990s world trade grew at 6.4 percent a year, and East Asian exports at more than twice that rate. This trade is being conducted in an increasingly liberal environment, with significantly lower tariff and nontariff barriers and with many new players, such as China, Russia, and other formerly communist countries. Advances in information and logistics technologies—key elements of the knowledge economy—have eroded traditional advantages of location. And perhaps the greatest facilitator of world trade in goods and services has been the vastly increased mobility of international finance, which creates new production capacity quickly in new locations.

Firms increasingly confront a more competitive trading and production environment, requiring continued vigilance to stay ahead of rivals. So it is not surprising that competitiveness has become a major preoccupation of national, regional, and local governments worldwide (see Box 1.1 for several views on competitiveness). Even rich countries that score high on various measures of competitiveness need to stay vigilant, as the U.S. Council on Competitiveness (1999: 5) understands:

> Despite eight consecutive years of economic expansion, the United States could lose its status as the world's pre-eminent innovator country in the next decade if current national policy and investment choices continue unchanged. . . . [This] may seem like an alarmist message. Yet the moment of greatest apparent success can be a country's moment of greatest vulnerability.

It is useful at the outset of a discussion of competitiveness to clarify the relationship between competitiveness and the knowledge economy. Competitiveness can be seen as a more focused component of the much broader knowledge economy. Fueled by electronic information technology, the knowledge economy draws on international best practice in policy, institution design, and logistics to lower the costs of producing and delivering goods. That allows firms to subcontract many more processes than

1

Box 1.1 Views on Competitiveness

"We should be a knowledge economy where the basis for competitiveness will be the capabilities and intellectual capital to absorb, process and apply knowledge. We should have a strong technological capability and a vibrant entrepreneurial culture that thrives on creativity, nimbleness and good sense. To develop into a knowledge economy, Singapore should be an open cosmopolitan society, attractive to global talent and connected with other global knowledge nodes. There should be a critical mass of Singaporeans who are risk-taking entrepreneurs, innovators and arbitrageurs. Together with the global talent, they will move Singapore ahead in the Information Age."

—Singapore's Competitiveness Vision, Committee on Singapore's Competitiveness, 11/1998

"Competitiveness has emerged as the preeminent issue in many nations. Achieving global competitiveness calls for a nation to upgrade its exports. Competitiveness also requires a nation's government and companies to have a shared vision about what competitiveness is and how it can be achieved. Competitiveness is *not* a simple macroeconomic adjustment, a favorable exchange rate, a positive trade balance, industrial subsidies, or a low inflation rate. Rather, competitiveness is the ability to achieve high productivity, relying on an innovative deployment of human resources, capital and physical assets. Competitiveness is the capacity to create value for increasingly sophisticated consumers who are willing to pay premium prices for the improved value that they perceive."

—The Monitor Company, 12/1997

"Improving competitiveness is central to raising the underlying rate of growth of the economy and enhancing living standards. Achieving this means removing the impediments to investments in machines, people and ideas and improving the efficiency with which resources are used throughout the economy, not just in those sectors directly involved in international trade. It means giving people the freedom to grasp new opportunities. It involves benchmarking all our activities against the best of our competitors to see how well we are doing compared to them and what we can learn from them.

"The need to improve our competitiveness is not imposed by Government, but by changes in the world economy. Improving competitiveness is not about driving down living standards. It is about creating a high skills, high productivity and therefore high wage economy where enterprise can flourish and where we can find opportunities rather than threats in changes we cannot avoid."

—U.K. Government, third competitiveness White Paper, U.K. Cabinet Committee, 1996

(Box continues on the following page.)

Box 1.1 (continued)

"'Competitiveness' is a growth industry. Presidents and prime ministers vow to improve it, legislators debate it, economists measure it, and editors feature it. In this context, the notion of competitiveness is typically couched in terms of one country or trading bloc versus another, and the animating question is whether country X is losing or 'surrendering' its competitiveness to country Y. When the unit of analysis is a firm rather than a country, the issue of competitiveness revolves around relative competitive position and competitive advantage. In this view, competitiveness comes from a 'defensible' market position and 'sustainable' competitive advantages. We believe the country versus country formulation of the competitiveness challenge is substantially inaccurate, and the 'position and advantage' formulation is incomplete."
—*Competing for the Future, Gary Hamel and C.K. Prahalad, 1994*

"Competitiveness is the ability to produce goods and services to meet the test of international markets while simultaneously maintaining and expanding the real incomes of citizens."
—*U.S. Presidential Commission on Industrial Competitiveness, 1993*

---

before and may well serve to redefine the concept of the firm. These aspects of the knowledge economy fall within the purview of competitiveness. However, the knowledge economy is sometimes amplified to include broader societal objectives such as broad-based participation in economic decisionmaking and cost-effective delivery of social services to the poor. These aspects of the knowledge economy are beyond the scope of competitiveness as discussed here.

Competitiveness imparts flexibility to firms, enabling them to respond to new opportunities and adjust quickly to changing market situations. However, building competitiveness or flexibility is a complicated process, and almost any improvement—expanding education or health, upgrading physical infrastructure, or reducing corruption in public agencies—is likely to enhance the business environment in which firms operate, in turn sharpening the competitiveness of firms operating in that environment.

The potential for public policy to affect competitiveness raises several important questions. To what extent is the behavior of firms influenced by the policy environment? Which public policies have proven effective in building systemic competitiveness? How should policymakers define priorities and benchmark performance? This paper develops a framework for examining some of these questions. Before developing a framework, however, it helps to consider various measures of competitiveness.

Measuring Competitiveness

In designing a reform agenda for building competitiveness, it is important to be able to benchmark performance across nations. Several variables have been used as indicators of competitiveness. Some of the most popular are discussed here: national competitiveness scores, foreign direct investment inflows, unit labor costs, and exports.

Indices of National Competitiveness

Two popular sources of national competitiveness scores are the *Global Competitiveness Report*, prepared by the World Economic Forum, and the *World Competitiveness Yearbook*, prepared by the International Institute for Management Development (details, Box 1.2).

The *Global Competitiveness Report* formulates an index of economic indicators correlated with medium- to long-term economic growth. The index combines data on a country's economic performance (trade, technological capacity, infrastructure, regulatory framework) with qualitative survey data from business executives on their perceptions of the business environment in the countries in which they operate.

The *World Competitiveness Yearbook* measures and compares how well countries are providing an environment that sustains the domestic and global competitiveness of the firms operating in their borders. It also uses

---

Box 1.2 Methodology of Two National Competitiveness Indicators

Global Competitiveness Report. The index is constructed from subindexes for the following factors:

- *Openness:* the openness of an economy to foreign trade, investment, foreign direct investment, and financial flows; exchange rate policy; and ease of exporting.
- *Government:* the role of the state in the economy, including the burden of government expenditures, rates of public saving, marginal tax rates, and the competence of the civil service.
- *Finance:* how efficiently financial intermediaries channel savings into productive investment, the perceived stability and solvency of key financial institutions, levels of national saving and investment, and credit ratings by outside observers.

(Box continues on the following page.)

Box 1.2 (continued)

- *Infrastructure:* the quality of roads, railways, ports, telecommunications; cost of air transportation; and overall infrastructure investment.
- *Technology:* computer use, spread of new technologies, ability of the economy to absorb new technologies, and the level and quality of research and development.
- *Management:* management quality, marketing, staff training, and motivational practices.
- *Labor:* efficiency and competitiveness of the domestic labor market.
- *Institutions:* extent of business competition, quality of legal institutions and practices, extent of corruption, and vulnerability to organized crime.

The index is a weighted average of the subindexes, with more weight given to quantitative data than to survey data: openness, 1/6; government, 1/6; finance, 1/6; infrastructure, 1/9; technology, 1/9; management, 1/18; labor, 1/6; and institutions, 1/18.

World Competitiveness Yearbook. The index is built on the following input factors:

- *Domestic economy:* macroeconomic conditions (value-added productivity, investment, savings, final consumption, sectoral performance, cost of living, forecasts).
- *Internationalization:* extent of a country's participation in international trade and investment (current account balance, exports and imports of goods and services, exchange rate, portfolio investments, foreign direct investment, state involvement, justice and security).
- *Finance:* performance of capital markets and quality of financial services (cost and availability of capital, stock markets dynamism, banking sector efficiency).
- *Infrastructure:* how adequately natural, technical, and communication resources serve the basic needs of business.
- *Management:* how innovatively, profitably, and responsibly companies are managed (productivity, labor costs, corporate performance, management efficiency, corporate culture).
- *Science and technology:* the scientific and technological capacity of the country (R&D expenditures, R&D personnel, technology management, scientific environment, intellectual property).
- *People:* availability and qualifications of human resources (population characteristics, labor force characteristics, employment, unemployment, education structures, quality of life, attitudes and values).

Each of the statistical data has the weight of 1. Weights for the survey-based data are computed to ensure that they have an overall weight of one-third, to temper the effect of volatile opinions.

two types of data to capture quantifiable and qualitative information. It obtains statistical indicator data from international and regional organizations, private institutions, and national institutes. And, through an in-depth questionnaire, it gathers qualitative data from top executives and middle management, who are asked to evaluate the current and future competitiveness of the country in which they operate.

Although the notion of national competitiveness scores is appealing, some cautionary comments on their construction and use are worth bearing in mind. First, a great deal of subjectivity is built into the scores through the interview-based methodologies, and the weights given to the qualitative attributes are not disclosed. Second, the scope of the measures is so broad that the computation of competitiveness includes variables with no clear causal relationship to competitiveness, but only a simple correlation (for example, stringent environmental regulations; see Lall 2001 for a discussion of the World Economic Forum's methodology). This lack of separation between the determinants of competitiveness and their outcomes is a serious flaw that the framework developed in this chapter tries to redress.

Foreign Direct Investment

Foreign direct investment (FDI), a measure of the investment underlying international production, is a reasonable measure of national competitiveness because the factors that make a country attractive for inward FDI are similar to those that determine its competitiveness. Countries compete for FDI flows. While most FDI inflows go to industrialized countries, among developing countries those in Latin America and East Asia dominate, as one would expect (Table 1.1).

Global flows increased fourfold between 1993 and 1999, and FDI stock accounted for 14 percent of world gross domestic product (GDP) in 1998, almost triple the 5 percent of 1980. The upward trend in all indicators of international production, in absolute terms as well as relative to various macroeconomic indicators, suggests that international production is becoming more prominent in the world economy.

While it is true that FDI flows into countries that possess the fundamentals of a competitive production structure, such as skilled labor, sound business laws, and good logistics, these are not the only factors affecting FDI inflows. For instance, the Republic of Korea, which possesses a sound production and innovation base, chose not to welcome FDI, unlike most other East Asian economies. China, on the other hand, based on the strength of its market size, has taken explicit measures to attract FDI and succeeded.

A final cautionary note about using FDI flows as indicators of competitiveness: FDI proved far more resilient during the East Asia financial crisis of 1997–98, staying reasonably stable, than did portfolio equity and debt

Table 1.1 Distribution of Foreign Direct Investment Inflows by Income Level and Region, 1981–85 and 1993–97

	1981–85				1993–97			
	Average value (US$ millions)	Distribution in world (percent)	Distribution in developing countries (percent)	Per capita (US$)	Average value (US$ millions)	Distribution in world (percent)	Distribution in developing countries (percent)	Per capita (US$)
World	56,375.4	100.0	n.a.	13.3	314,045.6	100.0	n.a.	63.4
Industrialized	42,541.8	75.5	n.a.	54.8	199,982.5	63.7	n.a.	241.6
Transition	—	—	n.a.	n.a.	9,597.5	3.1	n.a.	40.8
Developing	13,833.6	24.5	100.0	4.3	104,465.6	33.3	100.0	26.9
Income level								
High- and upper-middle	9,676.4	17.2	69.9	25.1	43,785.2	13.9	41.9	95.0
Lower-middle	2,505.2	4.4	18.1	4.8	18,280.0	5.8	17.5	28.5
Low	1,652.0	2.9	11.9	0.7	42,400.4	13.5	40.6	15.2
Low, excluding China and India	657.7	1.2	4.8	1.3	2,945.5	0.9	2.8	4.5
Region								
East Asia, including China	6,038.5	10.7	43.7	4.3	64,377.9	20.5	61.6	39.7
East Asia, excluding China	5,104.3	9.1	36.9	14.5	26,565.0	8.5	25.4	63.3
South Asia	196.6	0.3	1.4	0.2	2,522.9	0.8	2.4	2.1
Latin America and Caribbean, including Mexico	4,091.1	7.3	29.6	11.1	31,291.1	10.0	29.9	70.4

(Table continues on the following page.)

Table 1.1 (continued)

	1981–85				1993–97			
	Average value (US$ millions)	Distribution in world (percent)	Distribution in developing countries (percent)	Per capita (US$)	Average value (US$ millions)	Distribution in world (percent)	Distribution in developing countries (percent)	Per capita (US$)
Latin America and Caribbean, excluding Mexico	3,194.2	5.7	23.1	10.8	24,485.0	7.8	23.4	69.3
Sub-Saharan Africa, including South Africa	508.7	0.9	3.7	1.7	3,155.0	1.0	3.0	8.2
Sub-Saharan Africa, excluding South Africa	501.7	0.9	3.6	1.9	1,822.4	0.6	1.7	5.3
Middle East and North Africa	2,998.7	5.3	21.7	16.9	3,196.6	1.0	3.1	14.1

n.a. Not applicable.
— Not available.
Note: Annual averages calculated for available data; data were not available for Bahrain, Tanzania, and Yemen.
Source: World Bank, 2000; UNCTAD, 1995 and 1999; and national statistics.

flows, which suffered large reversals. The same was true during the Mexican crisis in 1994–95 and the Latin America debt crisis of the 1980s. These trends point to the relative inelasticity of this measure to changes in competitiveness in the short run. However, FDI is a good indicator of long-term or structural competitiveness of a country.

Unit Labor Costs

Unit labor cost—the cost of the labor input required to produce one unit of output, or the ratio of hourly compensation to labor productivity—is another commonly used indicator of competitiveness. As productivity rises, the labor input needed to produce a unit of output falls. An increase in productivity can offset an increase in compensation per hour in its effect on unit labor costs. Note for instance that Basel, Switzerland, with one of the highest wage rates in the world, has one of the lowest unit labor costs in the industrial world because of very high labor productivity.

So that costs can be compared across countries, unit labor costs need to be converted using a common denominator, such as the exchange rate. The variables required to calculate unit labor costs—wages, product prices, output, and exchange rates—embody both the micro and macro elements of the economy, and this characteristic gives unit labor costs a unique place among competitiveness indicators. For the countries shown in Table 1.2, unit labor costs declined during 1995–2000 in all countries except the United Kingdom; Belgium and the Republic of Korea registered the largest declines. However, calculating unit labor costs is far from easy given the sensitivities associated with obtaining accurate wage information and the difficulty of finding comparable baskets of goods across nations. Hence it is nearly impossible to get accurate unit labor costs for most developing countries.

Table 1.2 Index of Unit Labor Costs in Manufacturing, 1992 US$

Country	1995	1996	1997	1998	1999	2000
United States	94.8	93.5	91.9	92.7	90.4	89.9
Japan	131.7	109.6	97.7	92.4	102.4	102.5
Korea, Rep. of	122.3	123.3	95.8	63.7	65.4	67
Taiwan (China)	96.9	93	87.4	76	76.9	77.7
Belgium	105.2	98.3	81.2	80	76.9	66.4
United Kingdom	91.9	93	99.8	104.9	102.9	95.2

Source: Computed from national statistical office reports.

Export Performance

Success in export markets, measured by rising market shares, is an indicator of an economy's level of global integration. The structure of manufactured exports reflects a country's accumulated production, management and technological capabilities, industrial structure, business culture, policy framework, and institutions. Short-term export performance, of course, is also affected by the quality of macroeconomic management, especially exchange rates.

However, success in export markets needs to be interpreted carefully. For instance, the loss of some market shares in trade may not signify loss of overall competitiveness if there is a rising share of other products, signaling a move up the value chain (say, from garments to textile machinery). A proper picture of competitiveness requires specifying the relevant market shares, the causes of changes in shares, and the changes that are desirable for national welfare.

A useful extension of the simple export market share indicator is market positioning. In a matrix of share of production in world trade and share of exports in world trade, market position relates product-level market shares to the dynamism of exported products in world trade to show how a country is placed for growth in world markets (Table 1.3). A country's firms and industries are considered "competitive" in products in which their market shares are increasing. An export product is considered "dynamic" in world trade if its market share is growing faster than the average for all products.

The ideal market position is to have the highest share of exports as "rising stars," indicating that the country is gaining market share in fast-growing products. "Lost opportunity," the loss of market share in dynamic products, is the least desirable. "Falling stars" are also undesirable, although less so than lost opportunity, since market shares are rising but not in dynamic products. Finally, "retreat" may be undesirable, or it may be desirable if the movement is away from stagnant products and toward growth

Table 1.3 Matrix of Market Positioning

	Share of product in world trade	
Share of country's export in world trade	RISING (Dynamic)	FALLING (Stagnant)
RISING (competitive)	Rising stars	Falling stars
FALLING (noncompetitive)	Lost opportunity	Retreat

Source: Computer using TradeCan 2000.

in dynamic products. The rationale for applying this matrix approach is that competitive structures are difficult to change quickly, and the ability to adapt is unevenly distributed.

This analysis can also be extended in several directions. For instance, tracking the ratio of dynamic commodities to stagnant commodities exported could provide a measure of the responsiveness or adaptability of the export structure over time. Increasing numbers would reveal a strategic move in the right direction. An example for selected East Asian economies shows an interesting pattern (Table 1.4). Significant changes in the adaptability index of East Asian economies were occurring before the financial crisis of 1997, signaling a weakening of the underlying fundamentals of competitiveness, which ultimately led to the crisis.

This volume contains a paper on FDI and its determinants (Chapter 2). Further work on other competitiveness measures such as exports and unit labor costs is ongoing and will be reported separately.

A Framework for Examining Competitiveness

As these measures convey, a host of factors influences competitiveness. While public policy can facilitate the move toward the competitive frontier for firms, too often the reform agenda pursued is broad and ad hoc. The analytical framework developed here is designed to serve as a tool for diagnosing constraints to competitiveness and for prioritizing the reform agenda. It addresses two main questions: First, what makes firms worry about their competitiveness—what are the incentives facing firm managers that motivate them to operate on their competitive edge? These incentives can be seen as the demand side of competitiveness. Second, what are the fac-

Table 1.4 Adaptability Index: Ratio of Percentage of Exports in Dynamic Commodities to Percentage of Exports in Stagnant Commodities

Country	1992	1993	1994	1995	1996	1997	1998
China	2.41	0.37	0.43	0.30	0.33	1.95	2.02
Indonesia	0.65	0.27	0.33	0.51	0.52	0.54	0.59
Korea, Rep. of	1.74	1.06	1.28	0.85	0.94	2.00	2.02
Malaysia	2.17	1.55	1.88	0.98	1.15	3.19	3.44
Philippines	2.20	1.02	1.24	0.96	1.30	4.63	5.15
Thailand	2.08	0.73	0.88	0.61	0.69	1.47	1.55
Vietnam	0.84	0.21	0.26	0.34	0.36	0.57	0.60

Source: Computed using TradeCan 2000.

tors that facilitate the movement of motivated firms to their frontiers? These can be seen as the supply side of competitiveness.

Demand-Side Determinants: What Forces Firms to Worry about Competitiveness?

Four factors are key to strengthening the incentive regime that motivates firm owners and managers to improve competitiveness: strong shareholder rights and corporate governance, an adequate competition policy, a sound and prudent financial sector, and a balanced bankruptcy and secured lending regime (Table 1.5). In addition (but not discussed here), sound macroeconomic management is crucial to competitiveness. A realistic and stable exchange rate, low inflation, and low interest rates are all essential for ensuring that firms make the correct investment decisions and remain on their competitive edge.

STRONG SHAREHOLDER RIGHTS AND SOUND CORPORATE GOVERNANCE. One cause of the severe corporate distress in East Asia that led to the financial crisis of 1997 was excessive reliance on debt finance, a reflection of weak corporate governance (see Chapter 3). Inadequate disclosure and failure to protect minority shareholder rights, two manifestations of weak corporate governance, repressed the development of equity markets.

Transparency and *disclosure of corporate balance sheets* are powerful tools enabling shareholders to discipline management. To be effective, these tools depend on reliable accounting standards and auditing practices. Improving practices to bring them to the high levels attained by some countries in the region was an important component of the structural

Table 1.5 Who Motivates Firms to Become Competitive—and How?

Who	How
Shareholders	Auditing and accounting standards, code of ethics, disclosure rules, minority shareholders' rights.
Competitors	Competition law and policy, antitrust laws dealing with treatment of mergers, unilateral behavior of powerful corporations, horizontal agreements, vertical restraints, privatization, deregulation.
Bank supervisors	Prudential and regulatory standards, other financial institution supervision practices.
Creditors	Bankruptcy and secured lending regime, debtor-creditor relations, voting rules for institutional investors.

Source: Compiled by authors.

reform program implemented in response to the crisis. For example, on an index of accounting standards, Korea scored 64 and Thailand 62, while Malaysia scored 78 and Singapore 76, much closer to international best practice in 1998 (Dyck 2000).

The second pillar of corporate governance is *protection of minority shareholder rights*, to enable ordinary shareholders to take corrective action when firms perform poorly. On an index of minority shareholder rights developed by La Porta and others (1998), Thailand scored 3 out of a maximum score of 6, while Malaysia and the Philippines scored 4 each. For legal protections to be effective, strong judicial enforcement is also important. The regionwide average ranking on judicial enforcement is 5.4 (10 is the highest score), with Malaysia scoring a high 7.7, the Philippines a low 4.1, and Thailand a middling 5.9 (La Porta and others, 1998). As Chapter 3 argues, much remains to be done in the region to improve transparency and disclosure and to strengthen minority shareholder rights. The improvement in corporate governance thus achieved will hasten the development of equity markets, increase scrutiny of management by shareholders, and improve competitiveness.

ADEQUATE COMPETITION POLICY. Competition policy seeks to strengthen international competitiveness by exposing firms to competition (contestability) in the internal market (see Chapter 4). Contestability is impeded by hurdles such as licensing agreements, tariff and quota protection, and collusive behavior (resulting in price fixing). Such entry barriers restrict competition internally and result in a misallocation of resources that retards international competitiveness (see Box 1.3 for rationale for and approaches to competition policy).

The removal of quotas and licensing requirements is thus a critical aspect of competition law and policy. A related aspect is the prevention of collusive behavior such as horizontal agreements that result in price fixing and restrictions on output. Another focus is the prevention of abuse of market position by large dominant firms through predatory pricing intended to drive out competitors. These preventive measures are referred to as "conduct provisions" in competition law. Their enforcement entails the ability to distinguish practices that enable a firm to dominate the market through superior competitive performance from those that result in market dominance through collusive behavior.

Competition policy also seeks to streamline regulation of mergers, acquisitions, and joint ventures. Admittedly, such arrangements can discourage new entrants by increasing market concentration, but they also enhance efficiency by allowing large merged firms to enjoy scale economies or by replacing inefficient, entrenched management. Competition authorities conduct technical and economic analysis before permitting mergers, acqui-

Box 1.3 Rationale for Competition Policy and Various Approaches

The *structuralist school* emphasizes the interaction between market structure and collusive and exclusionary business practices that enable firms to exercise market power and persistently earn excess profits. Firms operating in oligopolistic industries with large market shares are more likely to coordinate their pricing and output. Both allocative efficiency and improved income distribution are considered valid objectives of competition policy.

The *Chicago school* argues that collusion is difficult to practice in all but the most highly concentrated industries and is therefore not a serious problem. Where competition is restricted, collusion arises primarily because of barriers to entry created by government. Exclusionary practices of firms are motivated by the pursuit of economic efficiency. The Chicago school therefore favors a minimalist approach, restricting competition policy to the prevention of price-fixing agreements.

The *statist or industrial policy school* argues that the competitive market is an outdated economic institution. Because markets often fail to guide investments to industries that would generate high growth, governments must lead the market by identifying strategic industries. In these industries, closer integration of business and government is needed to ensure that firms are large enough to compete with foreign firms. Competition policy should be set aside, since it impedes the ability of domestic firms to compete internationally.

These arguments and their variants have provided the rationale for different approaches to competition policy. Countries such as Canada, Colombia, Mexico, New Zealand, and the United States have emphasized economic efficiency; others such as Australia, France, India, and the United Kingdom have emphasized the impact of competition on the broad public interest, identifying issues relating to employment, diffusion of economic power, and regional economic development for the attention of their competition authorities.

sitions, or joint ventures in order to forecast their effect on competition in the industry and consequent effects on price and access for consumers.

Clearly, there is a fine distinction between what constitutes a legitimate practice in competition policy and what does not. Making that distinction requires sophisticated competition law, sound judicial practice, and a strong legal profession—institutions often lacking in developing countries—to prevent bureaucratic abuse in implementing competition policy. Although competition law may not be essential for fostering competition per se, it is increasingly being recognized that without it the economic and social ben-

efits in the form of greater economic efficiency, transparency, and participation cannot be safeguarded. One approach is to phase in competition
policy gradually. In the absence of a well-functioning legal system, specialized administrative bodies or oversight committees could be set up to
investigate competition-retarding behavior (Box 1.4). More complex investigations into abuse of market dominance may be phased in later, after the
skills of competition authorities have been upgraded.

**Box 1.4 Competition Policy Reform in the Republic of Korea
and Thailand**

Korea. Before the establishment of the Korean Fair Trade Commission
(KFTC) in 1981, Korean industry had become dominated by a few large
conglomerates (*chaebols*). The government picked strategic winners and
favored them through low interest loans, credit-rationing, and licensed
entry. Although KFTC examined a large number of cases, most were set-
tled with "warnings," and little effective competition was fostered
through the mid-1990s. In 1990, the powers of the KFTC were expanded,
and in 1994 it was declared an independent central agency under the
Office of the Prime Minister. In 1999 several key amendments were made
to the Fair Trade Act, broadening the grounds on which the KFTC could
intervene in mergers and acquisitions, weakening the ability of the *chaebols* to establish a legal basis for their control of affiliated companies, and
authorizing the KFTC to demand bank records of *chaebols* and their owners and owners' relatives in anticompetition investigations.
    Since these changes were introduced, the KFTC has been quite active.
Fines have increased, ranking high among countries with competition law
enforcement. However, the KFTC's monitoring of mergers and acquisitions activity has been limited to prenotification filings and compliance
with the statutory requirements for calculating the market-dominant position of merging firms. It has not (at least publicly) played a major role in
the "big deals" for restructuring the five largest conglomerates.

Thailand. During the financial crisis of the late 1990s, mergers and acquisitions became an important vehicle for corporate restructuring. As a
result, policymakers saw the importance of competition policy and made
a commitment to establish the institutional capacity needed to oversee
and enforce competition legislation, beginning with upgrading the skills
and resources of the Central Committee of the Department of Internal
Trade.

(Box continues on the following page.)

Box 1.4 (continued)

In 1999 Thailand replaced the Price Fixing and Anti-Monopoly Act of
1979 with the Business Competition Act and the Price of Goods and
Services Act. These laws are the legal basis for three substantive areas of
competition policy: prohibited practices; price fixing, collusion, and abuse
of dominant position; and mergers, acquisitions, joint ventures, and strategic alliances. An independent competition policy authority has been set up
to oversee implementation, to develop a strategy for disseminating information to the public, and to encourage a voluntary compliance strategy.

A SOUND AND PRUDENT FINANCIAL SECTOR. The recent crisis in East Asia
clearly demonstrates that weaknesses in the financial sector contribute to
poor investment decisions and eventual loss of international competitiveness. Financial sectors throughout the region expanded rapidly in the early
1990s, increasing three- to four-fold (measured by credit as share of GDP)
in a short time. The expansion took place within a weak legal and institutional framework for intervening in failing financial institutions. Regulatory
forbearance, outmoded prudential regulations, and lax supervision of financial institutions led to related-party lending, exposure to risky investment
projects, and rapid accumulation of foreign exchange risk by corporate borrowers. The problems were compounded by inadequate accounting and
disclosure practices that prevented the market from putting pressure on
borrowers to improve performance by reducing excessive risk taking.

The consequences of these unsound financial sector practices are well
known. Creditors lost the ability to ensure that funds deposited in financial
institutions were being intermediated to yield the highest sustainable
returns. A large share of investment funds was diverted into the nontradable sector, particularly into real estate and the retail business. Rapid expansion of these sectors raised wages across the board, but productivity did
not increase in the tradables sector because of underinvestment in upgrading skills and modernizing plant and equipment. The result was an increase
in unit labor costs throughout the region—another name for loss of international competitiveness.

Rapid expansion of bank lending also increased the vulnerability of companies to shocks and thus contributed indirectly to a loss of competitiveness.
The easy availability of credit encouraged companies to increase their debt-
equity ratios substantially and quickly. Much of this debt was denominated
in foreign currency. Thus when the correction eventually came, companies
were faced with substantially higher interest rates and drastically lower

currency values than when they had contracted the debt. Balance sheets deteriorated rapidly, and companies were forced to focus on day-to-day survival rather than on restructuring to improve international competitiveness.

Structural reform to strengthen the safety and soundness of the financial sector and thus restore international competitiveness was a top priority at the height of the crisis. Reforms focused on strengthening the financial sector through distress resolution (mergers, nationalizations, shutting down of financial institutions) and recapitalization (providing public funds for recapitalization, permitting foreign ownership of institutions). Strategies for asset resolution consisted mostly of setting up centralized agencies (Indonesia, Malaysia, and the Republic of Korea), although a more decentralized approach was also taken (Thailand). Financial sector regulation was improved by raising loan classification and provisioning standards to international best practice and ending policies of regulatory forbearance. Strengthening supervision also received attention. Thailand, in actions representative of reforms in the region, established policies for dealing with troubled financial institutions through prompt corrective action, consolidated supervision of financial institutions, and comprehensively reviewed the supervisory regime to strengthen the legal and regulatory framework and institutional capacity of the Bank of Thailand, the Ministry of Finance, and the Securities and Exchange Commission.

A BALANCED BANKRUPTCY AND SECURED LENDING REGIME. While the prudential, regulatory, and supervisory regimes are important for ensuring that financial institutions channel funds to their most profitable use, these need to be supplemented with a secured lending and bankruptcy regime that balances debtor and creditor interests in credit relations. If the legal regime weighs in too heavily on the side of creditors, too many bankruptcies may threaten entrepreneurial activity. If it is overprotective of debtors, they may escape the consequences of bad investment decisions, which could lower the efficiency of borrowed capital.

In a paper on the impact of the bankruptcy regime on debtor behavior, Claessens, Djankov, and Klapper (1999) find that debtor-friendly bankruptcy regimes, such as that of the United States, allow equity holders to retain management of firms during reorganization negotiations, while creditor-friendly regimes, such as that of the United Kingdom, allow creditors to replace management. Thus borrowers are more likely to be risk averse under the U.K.-style regime than under the U.S.-style regime.

For a proper assessment of risk aversion in different bankruptcy regimes, however, the costs inflicted by the bankruptcy process also need to be considered. Studies show that the U.S. bankruptcy regime exacts high personal costs from debtor-managers (their salaries and bonuses decline sharply), which results in a preference for out-of-court settlement over court-led bankruptcies. However, firms that opt for out-of-court resolution of distress remain highly leveraged and thus suffer further rounds of distress.

In many countries, banks own shares in debtor firms. In Japan, where such relations are widespread, studies find that while such relationships improve firms' access to credit, better enabling them to lower costs during distress, banks extract much higher rents from related firms than from other firms.

Using data on 4,569 firms worldwide and focusing on East Asia during the crisis, Claessens, Djankov, and Klapper (1999) report that the largest number of bankruptcies occurred in Korea, Malaysia, and Thailand and the smallest number in the Philippines, Singapore, and Taiwan (China). Nearly 60 percent of firms in Japan are bank related, compared with only 9 percent in Taiwan (China) and 14 percent in Korea and Singapore. Family-owned firms are dominant in Indonesia (72 percent), Hong Kong (China) (69 percent), and Thailand (64 percent), but not in Japan (14 percent).

In the East Asian region, creditor rights are stronger in countries with Anglo-Saxon and Germanic legal tradition, such as Malaysia, than in countries following the French tradition, such as the Philippines. Claessens, Djankov, and Klapper conclude that while family and bank relationships in East Asia provide insurance against possible bankruptcy, this protection is achieved at the expense of minority shareholders. Judicial efficacy is as important as the legal regime in increasing the likelihood of bankruptcy filing.

Supply-Side Determinants:
What Helps Firms Become More Competitive

The previous section examined the factors that motivate firms to increase their competitiveness. Once motivated, firms need to focus on developing the abilities that are key to improving competitiveness (Table 1.6):

- Ability to use, adapt, and innovate on existing technologies.
- Ability to attract, build, and retain appropriate human capital.
- Ability to manage logistics and improve the supply chain network.

ABILITY TO USE, ADAPT, AND INNOVATE ON EXISTING TECHNOLOGIES. Building innovative capabilities is not just about building high-tech plants or pushing out the frontiers of science. It also involves upgrading the technical infrastructure (information, equipment, management) at the manufacturing level in order to move to higher value-added activities (see Chapter 5). Firms can upgrade such capabilities by developing new technologies in-house, by purchasing new technologies domestically or abroad, or by rely-

Table 1.6 What Makes Firms Become Competitive—and How?

	How			
What	*Firm*	*Other firms*	*Institutions (public–private)*	*International agreements*
Adapt, absorb, modify technologies	Devote resources to R&D and adaptation	Spillovers, diffusion	Measurement, standards, testing, and quality; research and technology laboratories	Intellectual property protection agreements, WTO membership
Attract, build, and retain human capital	Devote resources to in-firm training	Spillovers (hiring workers trained by other firms), joint training arrangements	Educational institutions, skills development fund, vocational training institutes	Exchange and training agreements
Manage logistics and improve the supply chain	Devote resources to supplier and vendor development programs	Coordination among firms to integrate production and information systems	Physical infrastructure, quality, and certification institutions	Anti-trust, e-commerce agreements

Source: Compiled by authors.

ing on the natural process of technology diffusion—or by some combination of the three (Luthria 1999).

Developing new technologies entails a purposive commitment of time and human and physical resources. Formal research and development (R&D) is risky, expensive, and typically undertaken only by very large firms that can afford the costs to stay abreast of the latest technological advances, conduct market analysis to determine cost-benefit ratios of R&D, and secure exclusive rights to new breakthroughs.

Purchasing new technologies (either outright or by licensing) can be straightforward, as long as firms are aware of what technologies are appropriate and are knowledgeable about licensing issues. As with R&D, it is usually large firms that are able to identify useful new technology and negotiate purchase deals at competitive terms.

Relying on the natural process of technology diffusion is a more passive undertaking. Firms wait for new technologies to become common knowledge before incorporating them into production processes. While technology development in one industry can affect the production and cost structures of another through positive externalities, evidence shows that a firm's ability to adopt new technology hinges on its own absorptive abilities, which are closely tied to the firm's own engagement in some form of R&D (however loosely defined), casting doubt on claims that such upgrading can be truly "free" or "passive" (Cohen and Levinthal 1989).

Policies promoting technology upgrading. Public policy has tried to leverage each of these modes of technological learning. Most countries offer R&D tax incentives, enabling firms to write off a share of R&D expenditures. However, such incentives, although welcomed by industry at large, tend to be used by firms that would have undertaken R&D anyway, often resulting in a crowding-out of private research dollars. In other words, the incremental boost to R&D through such tax incentives is minimal.

Matching grant or loan schemes have worked better for encouraging smaller firms to engage in some sort of R&D. The technical and market feasibility of a grant or loan proposal is typically reviewed by a team of industry or university experts acting for the executing agency. Firms must invest some of their own resources to avoid moral hazard; amounts of 30–50 percent are common. Members of the review team may serve as mentors for the first few years. Such programs have been credited with introducing the concept of R&D to small and medium-size firms in China, Korea, and Turkey.

Public programs for remedying information asymmetries that prevent small firms from identifying appropriate technologies are common in industrialized nations. Older models of support took the form of large repositories of industry information, collected by government agencies or industry associations. The flaw in this design, however, was its assumption that firms would "know what they did not know" and would seek information. But by definition, information asymmetries prevent firms from identifying gaps in their knowledge.

Newer approaches in industrial countries rely on technology extension and demonstration centers to seek out and invite smaller firms to experiment with various technologies. These centers also offer mentoring and consultation on the compatibility of new technologies with existing processes and extensive training on negotiating licensing agreements. The manufacturing extension programs of the United States and the Steinbeis Foundation in Germany are examples of successful programs that have reduced the uncertainty and transactions cost associated with technology purchase for small and medium-size firms.

Public support for facilitating technology diffusion is based on the recognition that potential users, like potential developers, face uncertainty and learning costs, resulting in underinvestment in diffused technologies. And since diffusion is necessary for fueling a second round of innovation through modification and improvement, it is in society's collective interest to promote it. Successful policies have focused on bridging the cultural gap between the scientific community and private industry.

One popular method is technoparks and incubators. Physically locating firms, universities, and research institutes (or even large and small firms) together in the same campus or industrial park is thought to promote interaction among them. The reality, however, is that most technoparks and incubators have evolved into successful real estate operations with few joint innovation projects (Luger and Goldstein 1991). More successful has been a reduction in the public funding available to research and technology institutions, forcing them to generate revenue by selling their services to the private sector. While this is not easy for most research institutions, and a complete dependence on privately funded R&D is not advocated, forcing research institutes to generate some of their own revenue has improved coordination between technology generators and technology users, promoting the diffusion of newly developed technologies.

Finally, public interventions have supported protection of intellectual property rights. The relationship between intellectual property rights and the ability of firms to create, purchase, and modify technologies deserves special attention (see Chapter 9). The case for strong protection of intellectual property rights is based on the premise that innovative activity is constrained if the innovator cannot reap the rewards of innovation. The case for weak protection is based on the premise that knowledge is a public good and that technology diffusion is constrained by strong protection. Until recently, the level of protection for intellectual property rights varied with a country's level of development, with technology producers offering strong protection and technology users weak protection.

However, as investment flows seek global destinations, the ability of firms to protect their knowledge assets has become a critical determinant in their choice of destination. And all members of the World Trade Organization that are signatories to the Trade-Related Intellectual Property Rights (TRIPS) Agreement have committed to upgrading their intellectual property rights to industrial-world standards. The main concern in developing countries is that intellectual property rights reform is premature and will hurt their competitiveness, both in the short run by wiping out industries that have thrived on low-cost imitation and in the long run by reducing access to new knowledge. This concern has tended to polarize the debate on intellectual property rights, and progress by way of designing development-appropriate instruments has been slow.

To address these concerns, developing countries would need to engage all interested constituencies—government, the scientific community, supporters of private research, and industries that rely on access to foreign technologies—in open discussion to build consensus on the key concerns in each country. Implementation of TRIPS would need to be sequenced if tougher laws are not only to be put in place but also to be enforced. Policies should balance proprietary motives against access, efficiency, and distributional considerations using competition policy instruments or regulated pricing for certain essential technologies and products (pharmaceuticals). Appropriate mechanisms would also be needed to leverage indigenous knowledge (music and art, ancient medicine) and practices to create competitive advantage for domestic industry. These are new areas for public policy, and developing nations seek advice on how to proceed.

Measuring technological capability. Attempts are also under way to develop tools for benchmarking the capabilities of firms to search, process, adapt, develop, and improve technologies. Such benchmarking requires detailed knowledge of the technical, managerial, and information systems of each firm. While findings from such micro-level exercises are expected to become available soon, confidentiality concerns have prevented wide dissemination of results of this kind. Attempts to quantify the innovative capability of a country are more common, and results are readily available (*U.S. Council on Competitiveness* 1999; Roessner and others 2000).

ABILITY TO ATTRACT, BUILD, AND RETAIN APPROPRIATE HUMAN CAPITAL. A common policy response to skill shortages is to expand the capacity of vocational training institutes. However, skill requirements can change quickly with shifts in demand, so a supply-oriented response often results in mismatches between the skills supplied and those needed by industry. This mismatch can be avoided when training is provided by employers rather than public institutions, because employers know which new areas need improved skills (see Chapter 6).

The evidence on the links among training, output, wage growth, and productivity is strong. Using panel data, Bartel (1991) and Tan and Batra (1997) show that formal firm-led training has a positive and significant impact on firm-level productivity in several Latin American and East Asian economies. The likelihood of an employer providing training is closely related to firm size (large firms train more than small firms), education (workers with some education are more "trainable"), investments in new technology, use of quality control methods (which create the need for training), and foreign ownership (multinationals train more than local firms).

Market failures in training. Firms nonetheless may underinvest in training for fear that their trained employees will be poached by other firms. When training is only partly firm-specific and employers cannot shift any of the training costs to workers, high job mobility can prevent firms from recouping the costs of their investments. Also, imperfect information about training—its value, training pedagogy, and the skill requirements associated with using the new technology—can lead to underinvestment. This is especially pronounced among small firms that do not enjoy the scale economies associated with training and have poor access to credit.

Encouraging firms to train. The presence of market imperfections calls for selective government interventions to increase the extent and quality of training by private sector employers. Over the long to medium term, the most important intervention is the expansion and improvement of general education to enhance the trainability of future workers. In the short run, countries have adopted a variety of direct interventions such as subsidies for training through tax incentives, levies, and matching grants. When free-riding (poaching of workers) causes underinvestment in training, government can provide incentives for all firms to train and to share in the costs and benefits of training provided by other firms. A payroll levy on all firms is one such incentive. Several countries have had success with dissemination of best practices in training and technological know-how through industrial extension services, employer associations, and industry groups to encourage joint training programs and cost-sharing, and through matching training grants for firms or groups of firms.

ABILITY TO MANAGE LOGISTICS AND IMPROVE THE SUPPLY CHAIN NETWORK. As firms maintain only the functions closest to their core competencies and outsource much of the rest, coordination of outsourced activities becomes critical in determining the competitiveness of firms. Several entities are involved in synchronizing these activities—referred to as logistics or supply chain management (Box 1.5)—which promise significant efficiency gains for firms.

To put this process in a larger context, the logistics attributes of a national economy—the efficiency with which buy-sell-deliver transactions are completed—determine in large measure the share of globalization benefits an economy enjoys. Thus, for example, the ratio of aggregate transaction or logistics cost to GDP determines how efficiently working capital turns over—what productive return can be realized on investment in plant and equipment—and how efficiently markets clear at the national level. Capital committed to complete buy-sell-deliver transactions is not available for investment in productive fixed assets and human capital. Thus the difference in development potential is significant. A lean, high-service economy whose aggregate logistics-cost-to-GDP ratio is 10 percent will have 20 percent more capital free for productive reinvestment than a working-capital-intense, low-service economy whose aggregate-logistics-cost-to-GDP ratio is 30 percent.

Advantages for supply chain members. Individual suppliers, producers, and marketers who are associated through a supply chain coordinate their value-creating activities with one another and in the process create greater value than they can operating independently. In doing so, they distribute

---

Box 1.5 What Are Supply Chains?

Supply chains are institutional arrangements that link producers, processors, marketers, and distributors—often separated by time and space—that progressively add value to products as they pass along the chain. Supply chains distribute benefits and apportion risks among participants, thus enforcing internal mechanisms and developing chainwide incentives for ensuring timely production and delivery. Supply chains are thus the conduits through which:

- Products move from producers to consumers.
- Payments, credit, and working capital move from consumers to producers.
- Technology and advanced techniques are disseminated among producers, packagers, and processors.
- Ownership rights pass from producers to processors and ultimately to marketers.
- Information on current customer demand and product preferences at the retail level passes back from retailers to producers.

benefits and apportion risks among participants, thus enforcing internal mechanisms and developing chainwide incentives for ensuring timely production and delivery. Supply chains substitute intracorporate, contractual, or franchise affiliation for unaffiliated arm's-length transactions, enabling them to transfer risks among participants in the chain. Supply chains can progressively increase competitive advantage based on specialization among chain partners (see Chapter 8).

Well-designed supply chains can realize several kinds of captured value for their participants. Through quality control, they can ensure that exacting requirements (ecocertification, for example) of retail customers can be met or exceeded along each part of the chain. Through compression of the order-to-delivery cycle, improved demand forecasting, quicker supply response, and "strongest link" financing of the entire chain, supply chains can minimize the amount of working capital required to produce and deliver marketable products to consumers. Through optimal coordination of successive activities in the chain, supply chains can reduce product losses during storage and transportation. And through the sharing and application of knowledge among participants, including knowledge of customer preferences, market competition, product design, and process technology, supply chains can help to transfer knowledge efficiently.

Role of public policy. Clearly, the ability of firms to manage their supply chains is a function of the firms' organizational structure and their ability to manage all their activities well. However, the efficiency with which supply chains operate within an economy and the flexibility with which they can respond to new competition are also a consequence of public policy. Supply chains do not spontaneously emerge in all policy environments, but require fostering, supportive preconditions, and clarity of rules and regulations.

As a starting point, the efficiency of supply chains is strongly affected by the quality of infrastructure and its regulation—not only shipping and trucking infrastructure, but also the communication and information technology infrastructure that has become so critical in business transactions and is radically altering old ways of managing the supply chain. More is involved than the efficiencies gained from automation. Synchronized supply chains derive new value from being able to reach out to bigger markets though business-to business and business-to-consumer transactions conducted online and from performing mass customization of products and services to meet the needs of sophisticated new customers. (Chapter 7 presents an in-depth account of setting up e-commerce transactions.)

Antitrust laws are important for enabling suppliers to compete for a place on the supply chain in the first place, without being blocked by the collusive practices of large, dominant firms. Other aspects of the legal and regulatory regime, especially those related to liability, security, and insurance coverage of goods making their way through the supply chain, influence the length and breadth of the supply chain that can develop. In knowledge-intensive industries, protection of intellectual property rights can determine how easily certain functions can be contracted out.

Concluding Remarks

This chapter has presented an overview of the concept of competitiveness. It argues that policymakers need to distinguish between competitiveness outcomes and their determinants. A country's export market share (and various extensions of this simple concept) is a more useful measure of competitiveness outcome than the more broad-ranging indexes currently in vogue (such as those presented in the *Global Competitiveness Report* and the *World Competitiveness Yearbook),* which do not differentiate between outcomes and determinants.

The next step is to prioritize the many determinants of competitiveness to facilitate sensible policy intervention. Prioritizing requires specifying the relationships between determinants and outcomes and then quantifying them. The framework in this chapter focuses on specifying the relationships. It distinguishes between supply- and demand-side determinants. This distinction facilitates an understanding of the interplay of reform of the financial sector, bankruptcy and secured lending regime, corporate governance, and competition policy (the demand-side determinants) with the more well-known policy interventions for strengthening competitiveness by increasing skills training, technological innovation, information technology infrastructure, and the like (the supply-side determinants).

The discussion also sheds light on the type of interventions needed. The government has an indirect role in strengthening the demand side, which involves setting standards, laying down regulations, and ensuring enforcement. The government's role on the supply side is often more direct, requiring close interaction with the private sector in providing services (technology dissemination, skills deepening, and provision of telecommunication).

The chapter has not directly linked country performance on determinants of competitiveness with performance on competitiveness outcomes. This is work in progress.

The other chapters in this volume seek to deepen the understanding of the determinants of competitiveness. In much greater detail than in this chapter, the other chapters explore the links between competitiveness and its determinants and discuss best practice in implementing reform, highlighting the roles of the government and the private sector.

References

Bartel, Ann. 1991. "Productivity Gains from the Implementation of Employee Training Programs." Working Paper 3893. Cambridge, Mass.: National Bureau of Economic Research.

Cohen, Wesley, and Daniel Levinthal. 1989. "Innovation and Learning: The Two Faces of R&D." *Economic Journal* 99:569–96.

Claessens, Stijn, Simeon Djankov, and Leora Klapper. 1999. "Resolution of Corporate Distress: Evidence from East Asia's Financial Crisis." Policy Research Working Paper 2133. World Bank, Financial Economics Unit, Financial Sector Practice Department, Washington, D.C.

IMD International Institute for Management Development. 2000. *World Competitiveness Yearbook.* Lausanne, Switzerland.

Lall, Sanjaya. 2001. "Comparing National Competitive Performance: An Economic Analysis of World Economic Forum's Competitiveness Index." QEH Working Paper S61. Oxford University, Queen Elizabeth House, Oxford, U.K.

La Porta, Rafael, Florencio López-de-Silanes, Andrei Shleifer, and Robert Vishny. 1998. "Law and Finance." *Journal of Political Economy* 106: 1113–55.

Luger, Michael I., and Harvey A. Goldstein. 1991. *Technology in the Garden: Research Parks and Regional Economic Development.* Chapel Hill, N.C.: University of North Carolina Press.

Luthria, Manjula. 1999. A Framework for Enhancing Technological Learning among Firms. Occasional Paper No. 39, Private Sector Development Department, World Bank. Washington, D.C.

Roessner, David, Alan L. Porter, Nils Newman, and Xiao-Yin Jin. 2000. *A Comparison of Recent Assessments of the High-Tech Competitiveness of Nations.* Atlanta: Georgia Institute of Technology.

Tan, Hong, and Geeta Batra. 1997. "Technology and Firm Size-Wage Differentials in Colombia, Mexico, and Taiwan (China)." *World Bank Economic Review* 11:59–83.

UNCTAD (United Nations Conference on Trade and Development). 1995. *World Investment Report 1995.* Geneva.

————. 1999. *World Investment Report 1999*. Geneva.

U.S. Council on Competitiveness. 1999. *The New Challenge to America's Prosperity: Findings from the Innovation Index*. Washington, D.C.

World Bank. 2000. *World Development Indicators 2000*. Washington, D.C.

World Economic Forum and Center for International Development. 2000. *Global Competitiveness Report*. Cambridge, Mass.: Harvard University.

2

Foreign Direct Investment and Competitiveness

Sanjaya Lall

The factors that make a country attractive for inward foreign direct investment (FDI) also determine its competitiveness. Thus FDI is both a measure of the investment underlying international production and a reasonable measure of national competitiveness. With continuing technical change and globalization, the role of FDI is set to grow.

The Growing Importance of Foreign Direct Investment

The Asian financial crisis of 1997–98 demonstrated that FDI was the most stable element of foreign private financial flows: while FDI to the developing Asia region (including South Asia and the Middle East) was relatively stable after 1996 (a slight increase in 1997 was followed by a small drop in 1998), foreign bank lending and portfolio flows fell—lending precipitously. Following the shock, financial flows to developing countries as a whole rose by 15 percent in 1999, reaching a total of $198 billion (UNCTAD 2000). Inflows to developing Asia rose by 1 percent to reach $91 billion, reversing the decline of the previous year. This structural aspect of FDI has restored its prominence in development, which was somewhat diminished in the period when the massive value of short-term flows had attracted the attention of policymakers.

Perhaps more relevant to competitiveness is the growing role of FDI in upgrading productive activity under open trade regimes. International investors increasingly prefer cross-border mergers and acquisitions over greenfield investment. The great bulk of FDI in the industrial world—and an increasing share in the developing world—is now in the form of mergers and acquisitions, apparently because of the search for greater synergy and the pressure for rapid entry in the face of rising competition. The Organisation for Economic Co-operation and Development's *Science, Technology, and Industry Scoreboard* (OECD 1999: 11) notes:

> Since the second half of the 1980s, foreign direct investment has played a fundamental role in furthering industrial restructuring at the world

level. It most often takes the form of cross-border acquisitions of existing firms and is the fastest route to external growth for firms that seek to achieve "critical mass," increase rapidly their market share, rationalize their business or build up their technological potential and competitiveness.

East Asian countries are starting to welcome mergers and acquisitions as a way of involving local enterprises directly in globalization; the financial crisis has greatly increased their willingness to accept foreign acquisitions. Cross-border mergers and acquisitions in Asia (as a whole) reached an annual average of $20 billion during 1997–99, up from $7 billion during 1994–96. The most significant increase was in the five countries most affected by the financial crisis of 1997–98 (Indonesia, Malaysia, the Philippines, Republic of Korea, and Thailand). Their share of total cross-border mergers and acquisitions in Asia jumped to 70 percent in 1998, from 26 percent in 1996 (UNCTAD 2000), reaching a record $15 billion in 1999. Korea and Thailand accounted for the bulk of this activity.

There were large differences in FDI flows within East Asia. China retained its lead position in 1999, with inflows of about $40 billion, but this amount was a decline of around 8 percent from 1998. Several factors account for the drop. Domestic economic growth slowed. Several industries (such as garments and electrical appliances) had excess capacity because of past over-investment, while others faced increasing competition from neighboring low-wage countries. Outward FDI by Asian economies (significant investors in China) declined after the crisis. And the Chinese government was cautious in opening service industries to foreign investors. Despite these factors, China remains an attractive long-term location for FDI. Accession to the World Trade Organization (WTO) and further liberalization of services are likely to restore healthy growth in inflows.

FDI inflows to the East Asian Tigers—Hong Kong (China), Korea, and Taiwan (China)—increased by more than 40 percent from 1998 to 1999. The increase in Korea, nearly 55 percent (to $8.5 billion), was particularly striking. At four times the precrisis level, FDI inflows demonstrate the impact of policy liberalization on investors' perception of the robustness of Korea's economy. The government's public sector reforms and its need for financing led to large-scale privatization—another important attraction for foreign investors. Inflows to Taiwan (China) recovered to $2.4 billion from their exceptionally low level in 1998 of $222 million. Inflows to Hong Kong (China) increased slightly over the previous year to an estimated $8.1 billion in 1999.

FDI to Southeast Asia (ASEAN 10) decreased by 5–7 percent in 1999.[1] Thailand saw a fall of 15 percent, to $5.8 billion, in part due to the flattening of the wave of massive recapitalization in the banking sector that had

reached exceptional levels in 1998. However, Thai manufacturing continued to attract considerable FDI, to a significant extent through mergers and
acquisitions. Singapore continued to be the second largest FDI recipient in
East Asia with inflows increasing by 20 percent to $8.7 billion. Inflows to
Malaysia (at $3.8 billion) remained similar to those in 1998. The Philippines
suffered a decline, its overall inflows being low compared with most ASEAN
partners. Flows to Indonesia remained low as well, a sign of weak confidence in its economic reforms and political stability. Cambodia, Lao PDR,
Myanmar, and Vietnam also continued to suffer from the negative effects
of the crisis.

FDI is playing a growing role in the global economy. The direct involvement of multinational corporations is of increasing—often critical—importance to export competitiveness, especially for differentiated products with
global brand names and complex products for which technologies are
expensive, fast changing, or difficult to acquire. The significance of FDI is
even broader, however. Large multinational companies increasingly dominate innovation—the creation of new technologies and organizational
methods—which is at the core of competitiveness in all but the simplest
activities. Most of these multinational corporations originate in mature
industrial countries, and within these countries a small handful of companies account for the bulk of innovative effort. Thus, about 90 percent of
world expenditure on research and development (R&D) arises in OECD
countries: within this group, seven countries account for 90 percent of
R&D—the United States alone for 40 percent. In the United States just 50
firms (of a total of more than 41,000) account for nearly half of industry-
funded R&D. In smaller industrial countries the level of concentration is
even higher. In Switzerland just three firms account for more than 80 percent of national R&D. In the Netherlands four firms account for nearly 70
percent.

Access to new technologies thus involves access to the knowledge and
skills possessed by these technological leaders. These leaders are increasingly unwilling to part with their newest and most valuable technologies
without a substantial equity stake. Thus, FDI becomes the most important—often the only available—mode of obtaining leading-edge technologies. This is significant for the industrializing economies of East Asia, which
cannot now sustain competitiveness on the basis of low wages and simple
technologies. And FDI flows are growing faster than other economic aggregates such as national gross fixed capital formation (GFCF), world trade, and
gross domestic product (Figure 2.1).

Multinational corporations now dominate world trade: they handle
about two-thirds of visible trade (trade in merchandise), and their share is
growing. Their role is particularly large in products with significant
economies of scale in production, marketing, and innovation. About a third

Figure 2.1 Growth Rates of Foreign Direct Investment and Other Aggregates

Percent of growth

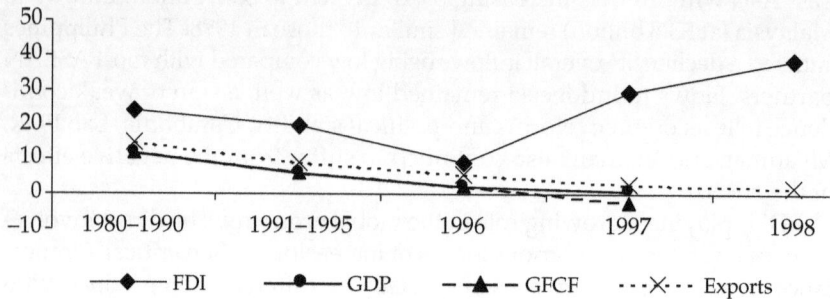

Source: UNCTAD 2000.

of the visible trade by multinational corporations is within multinational systems, between affiliates and parents or among affiliates. Such internalized trade contains the most dynamic form of exports today: integrated international production systems, with different functions or stages of production located in different countries and linked tightly together. Affiliates participating in such systems tend to realize enormous economies of scale, use the latest technologies, and deploy the latest skills and management methods. Examples of complex international systems in which developing countries are important are automobiles (mainly in Latin American countries like Argentina, Brazil, and Mexico) and electronics (Malaysia, Mexico, the Philippines, and Singapore). Some multinational corporations also locate nonproduction functions, such as accounting, engineering, R&D, and marketing, in different affiliates—high-value-added activities that yield considerable revenue and feed into manufacturing competitiveness and local capabilities.

Multinational corporations are often central to the export of technology-intensive products from local firms. Many such products are difficult to sell in foreign markets because of the need for expensive branding, distribution, and after-sales servicing. Some 60–70 percent of consumer electronic products made by Asian Tigers like Korea and Taiwan (China) are sold to multinational corporations on an original equipment manufacture (OEM) basis. Branded products sold by leading companies like GE, IBM, Compaq, Toshiba, and Sony are made by Korean and Taiwanese firms. In 1985 more than 40 percent of Korean exports were OEM. In 1989 about

50–60 percent of exports of video-cassette recorders and television sets and 80 percent of exports of personal computers by Korea were under OEM, and in 1990 some 70–80 percent of Korean electronics exports were under OEM. In the late 1990s OEM accounted for 40 percent of Samsung, 60 percent of LG, and 70 percent of Daewoo electronics exports.

Multinational corporations are also active in developing country exports of low-technology products in which branding, scale economies, distribution, or design is important, such as processed foods and high-quality clothing or footwear. The growth of export-oriented assembly operations throughout the developing world has been one of the major forces in increasing manufactured exports and boosting competitiveness.

Multinational corporations can be central to the restructuring and upgrading of the competitive capabilities of import-substituting activities geared to the protected markets of developing countries. Multinational corporations that already own the facilities are often better able than local firms to respond to liberalization by investing in new technologies and skills. They can also attract investment from their suppliers overseas or help local suppliers upgrade. This has been common in Latin America. Where local firms own the facilities, multinational corporations help them to upgrade through mergers and acquisitions. Although mergers and acquisitions by foreign firms are sometimes regarded with suspicion or resentment, they can salvage existing facilities that would not survive in a liberalized environment. In fact, with globalization and liberalization, international mergers and acquisitions now constitute the bulk of FDI flows, accounting for more than 80 percent of FDI in industrial countries and about a third in developing ones.

FDI in services is rising rapidly as traditionally homebound service providers (particularly in utilities) globalize their activities and take advantage of liberalization and privatization in their industries. The entry of multinational service corporations can bring rapid improvements in the productivity and efficiency of host economies, not only to their industries but also to their customers (many of which are important exporters).

In sum, FDI has become one of the major drivers of export competitiveness in the world today. In the developing world the highest shares of affiliates in manufactured exports are in Malaysia, the Philippines, and Singapore (over 70 percent in each), but multinational corporations also account for substantial shares of exports in China, Indonesia, and Thailand. In Latin America foreign affiliates account for 38 percent of exports in Argentina and 37 percent in Mexico (however, the foreign share of manufactured exports in Mexico is much higher, given the multinational corporations' dominant role in *maquiladora* exports). Box 2.1 shows the results of statistical analysis of the role of FDI in export performance in a large sample of developing and industrial countries.

Box 2.1   Statistical Analysis of Foreign Direct Investment and Export Performance

A regression analysis on 1995 exports—in total and by technological categories—for 72 industrial and developing countries shows a positive and significant relationship of foreign direct investment with export performance. The dependent variable is manufactured exports, measured in dollar values and per capita to account for country size. The explanatory variables tried were inward FDI, a skill index, research and development financed by productive enterprises, and per capita income. Three measures of FDI were tried: value of inflows during 1991–96, as a share of gross domestic investment (1985–90), and per capita for 1991–96. The results are similar, so only the per capita measure is shown here (Table 2.1). The skill index was derived from secondary and tertiary level school enrollments; other measures of skill, such as total years of education, were also tried. All these measures were highly correlated and gave very similar results; only the enrollment index is shown here.

FDI consistently has a positive and highly significant relationship with exports. The pattern of results does not vary with different combinations of variables and measures or with correction for heteroskedasticity. The relationship persists when some extreme observations are dropped. R&D also has a positive relationship with exports in most categories. Because skills are highly correlated with R&D, the results are not shown in the same regression; however, skills are statistically insignificant. Per capita income has a positive relationship with exports but is not shown because it is correlated with the other independent variables.

In general, the regressions have greater explanatory power for developing than for industrial countries. FDI plays a more important role in developing than industrial countries, while R&D shows higher significance in industrial than developing countries. The impact of FDI on developing country exports is particularly high in high-technology products, while for industrial country exports, it has the highest coefficient for medium-technology products. FDI also has high positive coefficients for resource-based and low-technology exports.

Influences on the share of complex (high- and medium-technology) products in manufactured exports are similar (Table 2.2). In general, the regressions are more significant (and highly so) for all countries together than for either developing or industrial countries and for developing than industrial countries, perhaps because structural differences are smaller between industrial countries. FDI—in all measures—has a positive and significant influence on the share of complex exports for all countries and for the developing country group, but it is not significant for industrial economies. Skills and R&D are tried in separate regressions, to minimize multicollinearity, and both have significant and positive effects.

(Box continues on the following page.)

Box 2.1 (continued)

While these exercises do not establish direct causal connection, they are plausible and suggestive. They confirm the (more scattered and qualitative) evidence that multinational corporations play a very important role in exports from developing countries, and their role is particularly large in complex and high-technology products (in developing countries, high-technology exports are strongly influenced by relocation of final assembly processes). The robustness and consistency of the results suggest that FDI is a real and positive factor in export performance.

Table 2.1 Determinants of Per Capita Manufactured Exports, 1995

Countries	Total manufactured exports	High-technology exports	Medium-technology exports	Low-technology exports	Resource-based exports
All (72)					
R&D	Positive***	Positive*	Positive***	Positive***	Positive**
Per capita FDI	Positive***	Positive***	Positive***	Positive***	Positive***
Adjusted R^2	0.83	0.66	0.71	0.66	0.70
Developing (47)					
R&D	Positive***	Positive***	Positive***	Positive***	Positive*
Per capita FDI	Positive***	Positive***	Positive***	Positive***	Positive***
Adjusted R^2	0.92	0.87	0.91	0.76	0.85
Industrial (25)					
R&D	Positive***	Positive***	Positive***	Insignificant	Insignificant
Per capita FDI	Positive***	Positive*	Positive***	Positive***	Positive***
Adjusted R^2	0.51	0.23	0.49	0.41	0.39

*Significant at the 10 percent level; **significant at the 5 percent level; ***significant at the 1 percent level.
Source: Author's computation and UNCTAD, World Investment Report.

Table 2.2 Determinants of Shares of High- and Medium-Technology Exports in Manufactured Exports, 1995

Variable	All countries (72)		Developing countries (47)		Industrial countries (25)	
	High technology	Medium technology	High technology	Medium technology	High technology	Medium technology
Skills	Positive***	—	Positive***	—	Positive*	—
R&D	—	Positive***	—	Positive***	—	Positive***
Per capita FDI	Positive*	Positive***	Positive**	Positive**	Insig-nificant	Insig-nificant
Adjusted R^2	0.37	0.52	0.19	0.22	0.04	0.55

— Not available.
*Significant at the 10 percent level; **significant at the 5 percent level; ***significant at the 1 percent level.
Source: See Table 2.1.

Traditional and New Determinants of the Location of Foreign Direct Investment

One of most important consequences of the new global environment is that decisions on where to locate FDI depend increasingly on economic factors that reflect underlying cost competitiveness rather than policy interventions that temporarily skew those decisions, such as offers of high levels of protection or tax incentives. Policy is not unimportant, however. On the contrary, because of increasing competition for FDI, policies affecting the investment environment and national competitiveness are more important than ever.

Traditional Determinants

Political and macroeconomic stability and a welcoming FDI regime are the most obvious traditional factors shaping decisions about where to locate FDI. Stability and predictability remain fundamental to attracting FDI under all conditions. So does a welcoming FDI regime, but the definition of "welcoming" is changing. With rising competition for FDI, more discriminating investors, and better communication and organizational technologies, a welcoming FDI regime increasingly means more than liberal rules set down on paper. It involves a proactive regime to promote and attract investment, with effective image building, low transaction costs for investors, equal treatment for all firms, and good support services.

Another important magnet for FDI is a large and dynamic domestic market. Traditionally, import protection reinforced the advantages of market size in attracting direct investment and was perhaps the most important single determinant of non-resource-based FDI. With liberalization the importance of this factor is declining. Even markets with relatively high protection are set to liberalize in the medium term, and many foreign investors base their plans on a freer market. Even with liberal trade regimes, however, a large and growing market remains a powerful draw. For instance, the formation of the European Economic Community some decades ago was a powerful incentive for foreign investors; similar trends are evident in Latin America. In manufacturing, a direct presence in a large, profitable market is considered important for preserving or increasing market share (the need is stronger when products have to be adapted to local tastes). For services the attractions of large markets are more evident, since services are largely nontradable.

The other important traditional determinants of FDI are exploitable natural resources and a cheap, disciplined, and skilled or trainable labor force. A cheap and skilled labor force has been the engine of growth of much recent export-oriented FDI in assembly activities in the developing world, both in low-technology activities like clothing and in final operations in high-technology activities like electronics. However, the nature, skills, adaptability, and working conditions of the labor force needed to sustain such FDI are changing (see next section on new determinants).

For export-oriented activity, access to major markets and quota availability in products like textiles and clothing under the Multi-Fiber Arrangement are also important determinants of FDI location. With the spread of trade liberalization, again, this sort of privilege will play less of a role. Good basic infrastructure (including well-run export processing zones) provides the necessary condition for competitive supply chain management.

New Determinants

The interaction of technical progress with liberalization and greater competition for FDI among host countries has led to significant changes in the factors affecting FDI location decisions. Decisions are now based more on the underlying competitiveness of economies.

Low-cost "raw" labor is being replaced by the need for qualified human capital able to cope with emerging technologies at world best-practice levels. While some investors still seek cheap, unskilled labor, they tend to be at the lowest end of the technological spectrum, with low sunk costs and a propensity to move as wages rise. These footloose investors offer very short-term benefits, and their spillover effects on creating new skills, technologies, and supply linkages are often negligible. Countries that seek more complex, sustainable, and high-quality FDI—with a long-term perspective and beneficial spillovers—must provide a growing and diverse supply of modern skills. The attractiveness of skills depends not just on the level of education, but also on the quality and relevance of the education, technical vocational training, and specialized postschool skill training and on the availability, quality, and responsiveness of employee training institutions. The quality of human resources in the broader sense also depends on work attitudes and practices and the impact of labor unions and regulations on hiring and firing workers.

Natural resources retain their significance for FDI. Resource-rich countries able to offer attractive and stable conditions for exploration and development are likely to continue to attract investors. But unless the economy can create the necessary skills and supplier bases, it will not be able to move up the value chain. For further development based on resources, therefore, other new determinants of FDI location come into play.

Competitiveness in the globalized world increasingly requires efficient logistics—modern infrastructure and supply-chain management. Efficient logistics in turn require rapid response, reliable and low-cost delivery, and instant communication. Basic infrastructure is no longer sufficient to attract FDI into many export-oriented activities. What is needed is state-of-the-art logistics structures with modern transport, telecommunication, data-linking, and other facilities. These in turn have to be supported by relevant consulting, facilitation, support, and maintenance services. Export processing zones and industrial estates have to meet increasingly stringent demands of facilities and services. Some developing countries have been able to develop such infrastructure and services, but many have not.

In many activities competitiveness depends on the proximity of supplier networks and clusters—efficient suppliers and service firms able to meet the needs of just-in-time production. Multinational corporations often draw follow-on investments by their established supplier and service firms (if other conditions are favorable), but countries with dynamic local firms (including small firms) have an advantage in attracting more FDI and greater local linkages.

Competitive operation often requires strong support institutions and technical services, including research and technical infrastructure. For host economies that want to stimulate R&D and technological spillovers from foreign investors, essential infrastructure facilities include effective quality assurance and testing bodies, measurement and calibration services, contract research, and technical extension help for small and medium-size firms. Research universities can help attract technology-based FDI and serve as focal points for disseminating new technological knowledge.

Countries vary substantially in their ability to promote FDI and serve investors. Countries with agencies able to project and improve the country's international image, attract the most promising investors at the national or firm level, provide efficient and honest services, and meet the emerging needs of investors are likely to attract the most FDI. Promotion works best when the legal regime is efficient, with fast dispute settlement mechanisms and equal treatment of foreign and local investors.

Because a substantial proportion of FDI flows to developing countries is in privatizations of state firms, particularly in infrastructure services, openness to FDI in privatization, utilities, and services is important. The ability to attract this investment demands appropriate rules and legal systems. The ability to benefit from it depends on the ability to regulate firms effectively and on a transparent, predictable basis.

A vital competitive tool in the FDI game is low transaction costs for investors, covering a wide range of bureaucratic interfaces with business: entry and exit rules, trade and labor market regulations, export and import requirements, location and environment regulations, tax payments, legal systems, and so on. Transaction costs vary enormously by country and depend not just on rules, but also on how they are implemented. The honesty, responsiveness, and skills of the bureaucracy in dealing with investors affect operational costs significantly; in a liberalizing world even small differences in such costs affect location decisions. Discriminating investors (especially those with valuable technologies) increasingly require locations with low, predictable, and competitive transaction costs.

Multinational corporations require great freedom in moving skilled personnel between locations with the least delay and hassle. They prefer, when possible, to use local personnel because the cost is lower. However, the pace of technical change and the need to keep close control of critical decisions in affiliates mean that firms need a highly mobile set of technical and managerial personnel across the globe (many come from other developing countries). To accommodate this need for personnel mobility, host governments have to relax traditional rules on the use of expatriate personnel, the issuing of work permits, and the setting up of facilities for foreign residents. Conducive and pleasant environments for families also figure increasingly in FDI attraction.

Policy Needs for Foreign Direct Investment

Technology flows across economies in many ways, disembodied and embodied. Its effective transfer and subsequent development depend on the channels of transfer and, increasingly, on local abilities to use them. With rapid change and a growing reliance on information, the abilities needed become more varied and skill intensive. Technical progress has led many channels of technology transfer to expand (and often become cheaper), although at the advanced end of the spectrum, it has made access more difficult. The rising costs of innovation, the spread of international production, and policy liberalization have increased the role of multinational corporations in all aspects of technology. This does not, however, mean that developing countries can simply open their doors to FDI and passively rely on it to transform their technological base. Deficiencies in technology markets mean that there can be an important role for policies in stimulating technological dynamism (see Chapter 5).

Different Strategic Approaches

East Asia's experience shows that two broad strategies are feasible, one relying on internalized technology inflows through FDI and the other on externalized technology inflows (see appendix for a statistical illustration of the strategies). The FDI-dependent strategy entails lower domestic technological effort. It is thus less risky and costly but can result in less development of domestic technological capabilities. One form of the internalization strategy involves considerable government intervention and targeting to promote technology upgrading. Singapore, for example, had the skill base and administrative capability to select technologies, target and bargain with multinational corporations, and handle incentives efficiently (Box 2.2). Another, more passive internalization strategy relies on market forces to attract FDI and technology transfers. This strategy requires a conducive environment for FDI and a cheap and trainable work force, but it has not resulted in comparable upgrading of innovative capabilities in host economies. The externalization strategy was practiced successfully mainly in Korea and Taiwan (China), which restricted inflows of FDI and promoted technology inflows in other forms while supporting domestic enterprises in mastering increasingly complex activities.

Korea's strategy relied less on FDI and more on direct government intervention to build technological capabilities in advanced and risky activities (Box 2.3). Although the government is now withdrawing from detailed industrial interventions, its early strategy clearly raised the technological capabilities of domestic firms to levels almost unmatched in other newly industrializing countries. The strategy was risky and costly, however, and very demanding of government skills and bureaucratic autonomy. No other country with restrictive strategies on FDI (of which there were many, like India) has achieved comparable competitive success in technology-intensive activities. Moreover, the accelerating pace of technical change and the spread of international production also make autonomous strategies more costly than before—one reason Korean firms are increasingly seeking international alliances to keep abreast of competitive demands. Needless to say,

Box 2.2 Singapore's Foreign Direct Investment Strategy

After a brief period of import substitution, Singapore switched to free trade. It pursued growth by aggressively seeking FDI and mobilizing domestic resources. It chose to deepen its industrial and export structure, using selective interventions to move from labor-intensive to capital-, skill-, and technology-intensive activities. Its technology acquisition policy was directed at acquiring and upgrading the most modern technologies in highly internalized forms. This strategy allowed the country to specialize in specific stages of production within global production systems, drawing on the flow of innovation generated by global firms and investing relatively little in its own innovative effort.

Singapore developed a highly efficient system of attracting and targeting multinational corporations. It developed a higher technical education structure targeted on industry, together with one of the best systems in the world for specialized worker-training. Some of the leading training centers were set up jointly with multinational corporations.

Its FDI policies were based on liberal entry and ownership conditions, easy access to expatriate skills, and generous incentives for the activities that it was seeking to promote. It set up the Economic Development Board (EDB) in 1961 to coordinate policy, offer incentives to guide foreign investors into targeted activities, acquire and create industrial estates to attract multinational corporations, and generally mastermind industrial policy. At times, it deliberately raised wages to accelerate technological upgrading, although in the mid-1980s a sharp rise in wages had to be modified to restore competitiveness. Over time, multinational corporations were drawn into the industrial policymaking process, and the EDB emerged as one of the world's most successful investment promotion agencies.

The public sector played an important role in launching and promoting activities chosen by the government, acting as a catalyst to private investment or entering areas that were too risky for the private sector. In recent years the government has also sought to increase linkages with local enterprises by promoting subcontracting and improving extension services. The government launched R&D centers to create new capabilities in the economy, which would later attract participation by multinational corporations.

The decisions of multinational corporations about what new technologies to bring into Singapore were strongly influenced by the incentive system, the provision of excellent infrastructure, and the direction offered by the Singapore government. Often it was the speed, efficiency, and flexibility of government response that gave Singapore the edge over competing host countries. Consider the boom in investment in offshore production by multinational corporations in the electronics industry in the 1970s and

(Box continues on the following page.)

Box 2.2 (continued)

the early 1980s. The Singapore government seized this opportunity by ensuring that enabling support industry, transport, communication infrastructure, and skill development programs were available to attract these industries. This concentration of resources helped Singapore achieve significant agglomeration economies and establish strong first-mover advantages in related industries, such as assembly support and precision engineering support. These supporting industries were actively promoted by the government.

As labor and land costs rose, the government encouraged multinational corporations to reconfigure their operations. A special program was launched to make Singapore attractive as a regional headquarters for multinational corporations and for marketing, distribution, service, and R&D centers to support manufacturing and sales operation in the region.

---

the current rules of the game also rule out many of the interventionist instruments used by Korea.

There are examples of innovative policies to promote local technology generation in the newly industrialized economies. Taiwan (China), for example, has mobilized local research consortiums to collaborate with multinational corporations in developing new technologies (Box 2.4).

The best policies on technology transfer by multinational corporations depend on the context: technology, government, and recipient enterprise capabilities and the learning environment. What is appropriate for high

---

Box 2.3 Foreign Direct Investment and Technology Development Strategies in the Republic of Korea

The Korean government combined selective import substitution with forceful export promotion, protecting and subsidizing targeted industries that were to form its future export advantage. To enter heavy industry, promote local R&D capabilities, and establish an international image for its exports, the government promoted the growth of giant local private firms, the *chaebol*, to spearhead its industrialization drive. Korean industry built up an impressive R&D capability by drawing extensively on foreign technology in forms that promoted local control. Thus the government encouraged Korean firms to obtain the latest equipment and technology, making Korea one of the largest importers of capital goods in the devel-

(Box continues on the following page.)

Box 2.3 (continued)

oping world. It encouraged the hiring of foreign experts to help resolve technical problems.

The government consistently sought to keep control in local hands, permitting FDI only when other means of accessing technology were unavailable. Foreign majority ownership was not permitted unless it was a condition of access to closely held technologies or needed to promote exports in internationally integrated activities. Some multinational corporations were induced to sell their equity to local partners once the technology transfer was complete. In the initial stages of development of important industries like electronics, however, multinational corporations played a major role in launching export-oriented assembly.

When the pace of technological upgrading of foreign affiliates was slower than desired, the government pushed local firms to acquire independent capabilities, ranging from mastery and improvement of imported technologies to absorption of foreign management practices and innovative R&D. The government supported technological effort in several ways. The Law for the Development of Specially Designated Research Institutes provided legal, financial, and tax incentives for private and public institutes in selected activities. Private R&D was directly promoted through incentives, including tax-exempt Technology Development Reserve funds and tax credits for R&D expenditures and for upgrading human capital related to research. The government also gave accelerated depreciation and tax incentives for investments in R&D facilities, and reduced duties and gave tax incentives on imported research equipment and other technology imports. The Korea Technology Development Corporation provided technology finance, and the Korea Technology Advancement Corporation helped firms to commercialize research results.

The government directly financed a large number of projects judged to be in the national strategic interest, including three R&D programs: the Designated R&D Program, the Industrial Technology Development Program, and the Highly Advanced National Project Program.

The main stimulus to the tremendous growth of industrial R&D was less the specific incentives to R&D than the overall incentive regime. It resulted in the creation of the *chaebol*, which provided a protected market in which to master complex technologies, minimized their reliance on FDI, and forced companies into international markets where competition ensured that they would have to invest in their own research capabilities. That is why, for instance, R&D spending by industry as a proportion of gross domestic product is 35 times higher in Korea than it is in Mexico (with roughly the same size of manufacturing value added), which has remained highly dependent on technology imports. At the same time these incentives may not have created sufficient innovative capabilities in *chaebol*, which excel more at implementing than creating state-of-the-art technologies.

Box 2.4 Technology Development through Alliance Formation in Taiwan (China)

In June 1995 in New York City, IBM unveiled its first personal computer (PC) based on the new PowerPC microprocessor, a product made by an alliance of IBM, Motorola, and Apple. One day later in Taipei a group of 30 firms from Taiwan (China) unveiled PowerPC-based products. They were the first non-U.S. firms to develop state-of-art products based on the new technology. The Taiwanese firms did not do this on their own. They were part of an innovation alliance, the Taiwan New PC Consortium, formed in 1993 by a government research institution—the Computing and Communications Laboratory (CCL)—to bring together firms from all parts of the information technology industry in Taiwan. Its specific purpose was to transfer, master, and diffuse the new PowerPC technology over the whole range of products—from PCs and peripherals to software and multimedia applications, as well as semiconductor manufacturers that would make their own versions of the new chip. The firms were small by international standards. CCL brought them together and negotiated on their behalf with IBM and Motorola.

This was just one instance of strategic alliance formation by the government of Taiwan to stimulate innovation and take industry to technological frontiers. The Industrial Technology Research Institute (ITRI) led in the formation of some 30 consortiums in the information technology industry during the 1990s. ITRI focused on products like laptop computers, high-definition television, videophones, laserfax, broadband communications, digital switching devices, and smart cards, helping firms move up the technology chain. ITRI identified the products, tapped channels of technology transfer, mobilized the firms, and handled the complex negotiations with industrial country firms, including intellectual property issues. Firms developed their own versions of the jointly developed core products and competed in final markets at home and abroad. Their size limited their ability do this on their own.

---

technology or a highly industrialized economy is not appropriate for a simple technology or a less developed country. The less developed the country and the lower its domestic capabilities, the more it needs FDI to overcome obstacles to technological mastery in relatively complex activities. Government capabilities are crucial. Experience shows that pervasive selective interventions, if pursued inefficiently, impose high penalties on the economy. In the emerging market setting, the approach most likely to succeed is a policy of attracting FDI, accompanied by incentives for upgrading and strong efforts to build domestic skills, technological capabilities, and institutions. This policy approach can be grouped into three categories: technology transfer, technology diffusion, and technology generation.

POLICIES ON TECHNOLOGY TRANSFER. The most important determinants of technology transfer are the level of skills and capabilities of the affiliate, its competitors and supplier network, and the competitive environment. The higher the level of local capabilities and the more competitive the environment, the better the quality of the initial transfer and the more rapid its upgrading. Multinational corporations invest in strengthening in-house skills and technical knowledge to the extent necessary to achieve efficient production but not necessarily to raise capabilities to the next level of technology. To achieve this, countries need policies that

- Change the competitive environment and incentives to promote the use of world-class technologies and management methods. Two of the most important reasons for technology upgrading by multinational corporations in Latin America have been liberalization of the trade regime and the provision of industry-specific incentives for export promotion. These moves have led to a regional rationalization of the industrial structure, with much greater specialization to reap scale economies and a massive upgrading of technologies.
- Improve the skill base and employee training. Policies should aim to raise the quality of the labor force outside the firm (general education policies) and encourage better training of employees within the firm. In-firm training generally suffers from what economists term "appropriability" problems: firms are reluctant to invest in providing transferable skills to employees who might leave the company. To overcome these problems governments can offer incentives for in-firm training or encourage firms to contribute to general training that benefits all firms (see Chapter 6). It is important to involve firms in decisions about the appropriate set of policies and skill needs (Box 2.5).
- Offer incentives to existing investors to move into more complex technologies and upgrade the technological functions undertaken locally. Policies should focus on upgrading all factor inputs that multinational corporations need (infrastructure, skills, information, and the like) and giving targeted incentives for launching new functions. The nature and level of incentives can be geared to the specific technological objectives, and can be designed in consultation with multinational corporations and according to successful experiences (Ireland and Singapore are good role models).
- Improve technology access for local enterprises, by providing information on foreign and local sources of technology.

Box 2.5 Human Capital Formation for Industrialization in Singapore

The Singapore government has invested heavily in creating high-level skills to drive the targeted upgrading of the industrial structure. The university system was expanded and directed toward the needs of the government's industrial policy; its specialization changed from social studies to technology and science. The government exercised tight control of curriculum content and quality.

The government also shaped the industrial training system, now considered one of the best in the world for high-technology production. The Vocational and Industrial Training Board (VITB) established an integrated training structure. Some 9 percent of the work force has passed through the system since its inception in 1979. The VITB administers several programs. The Full-Time Institutional Training Program provides broad-based preemployment skills training for school leavers. The Continuing Skills Training Program offers part-time courses and customized courses based on specific requests from companies. Continuing Education provides part-time classes to help working adults. The Training and Industry Program offers apprenticeships to school leavers and former members of the military service. Apprentices, working under the supervision of experienced and qualified personnel, acquire skills on the job. Off-the-job training includes theoretical lessons conducted at VITB training institutes or industry training centers. The government has collaborated with foreign enterprises to set up these centers, funding a large part of employee salaries during training in state-of-the-art manufacturing technologies, and has also worked jointly with foreign governments to provide technical training.

Under the Industry-Based Training Program, VITB assists employers in setting up skills training courses matched to their needs. VITB provides testing and certification of trainees and apprentices as well as trade tests for public candidates. In collaboration with industry, VITB certifies service skills in retailing, health care, and travel. Through grant schemes the National Productivity Board's Skills Development Fund created 405,621 training places in 1990. The fund has increased its impact on smaller firms through special efforts to make them aware of the training courses and of financial assistance schemes to help them finance their training needs and upgrade their operations. It has also introduced a Development Consultancy Scheme to provide grants to small and medium-size enterprises for short-term consultancy for management, technical know-how, business development, and personnel training.

Through the Training Voucher Scheme, which helps employers pay training fees for employees, the Skills Development Fund reached more than 3,000 new companies in 1990, many with 50 or fewer employees.

(Box continues on the following page.)

Box 2.5 (continued)

Initially, efforts focused on creating awareness among employers, with ad hoc reimbursement of courses. The policy was then refined to target in-plant training, and reimbursement increased to 90 percent of costs as an additional incentive. Further modifications were made to encourage the development of corporate training programs by paying grants in advance of expenses, thus reducing interest costs to firms.

The Economic Development Board continuously assesses emerging skill needs in consultation with leading enterprises and prepares specialized courses. For instance, in 1998, it offered courses on wafer fabrication, process operation and control, precision engineering, high-end digital media production, and computer networking. In 1991 it started an International Manpower Program to help Singapore-based companies attract skilled personnel from around the world. With its assistance some 2,500 professionals and 10,400 skilled workers and technicians were recruited in 1997.

POLICIES ON DIFFUSION. Technology diffusion by multinational corporations to vertically and horizontally linked enterprises depends greatly on the receptive capabilities of those enterprises and the competitive environment. In addition to the general measures discussed above, specific measures can strengthen linkages between multinational corporations and local suppliers, particularly small and medium-size firms. Such measures include

- Improving extension and training services to strengthen the capabilities of small and medium-size firms.
- Concentrating on particular areas or clusters of enterprises or on particular activities where spillover effects from multinational corporations are especially valuable.
- Offering incentives to multinational corporations to develop local suppliers. For example, under the Local Industries Upgrading Program, the Singapore government encourages multinational corporations to adopt a group of small firms and transfer technology and skills to them. It pays the salary of a full-time procurement expert to work with the adopted firms.
- Encouraging contract R&D with local research institutions and universities by reforming the institutions themselves to make them more industry oriented, strengthening their research capabilities, and perhaps underwriting part of the cost of approved research contracts.

- Setting up new research institutions on topics of special interest to multinational corporations. Between 1985 and 1995 Singapore set up nine research centers focusing on information technology, biotechnology, and electronics, to provide specialized training, develop precompetitive technologies, and provide services to companies.
- Encouraging technology alliances between local firms and multinational corporations by offering fiscal benefits for the cost of the research or the exploitation of its results.

POLICIES TO GENERATE R&D. In addition to the policies for encouraging technology transfer and the upgrading of affiliate functions, policies are needed to encourage local R&D. They include

- Offering fiscal benefits or grants to affiliates to handle an entire product for the parent company, from design to marketing. Such incentives can be effective where other capabilities are present in the host economy.
- Offering incentives for local R&D more generally, perhaps matching the level of the incentives to the nature of the technology and research undertaken. Advanced work in strategic areas such as information technology and producer electronics can, for instance, be given stronger than usual incentives.
- Improving industrial parks. Technology development requires first-class infrastructure, particularly advanced and low-cost telecommunications.
- Developing university research laboratories and research institutes and connecting them to multinational corporations and to companies in other countries that contract for such services. India has used this strategy very successfully.
- Providing tax incentives for R&D and duty-free importation and accelerated depreciation for research equipment.
- Accelerating technology generation by enforcing intellectual property rights, such as patents and trademarks (see Chapter 9). Protecting intellectual property will encourage technology generation not only by multinational corporations, but by domestic companies as well, but it might also impede efforts to "reverse-engineer" technology, a major source of technology for domestically owned firms in developing countries.
- Encouraging investment by foreign universities to accelerate technology transfer, dissemination, and generation and to increase the educational and skill levels of the labor force. Malaysia has followed this strategy.

Attracting and Promoting Foreign Direct Investment

Multinational corporations increasingly move their portfolios of mobile assets around the globe, seeking the best match in immobile assets. In the process, they also move some of the functions that create their ownership assets, such as R&D, training, and strategic management, within an internationally integrated production and marketing system.

SOURCES OF COMPETITIVE ADVANTAGE. With policy liberalization and technology change, the opening of markets creates new opportunities and challenges for multinational corporations and gives them a broader choice of modes for accessing those markets. It also makes them more selective in their choices of potential investment sites. The ability to provide competitive immobile assets—world-class infrastructure; skilled and productive labor; and an agglomeration of efficient suppliers, competitors, support institutions, and services—thus becomes a critical part of an FDI and competitiveness strategy for developing countries.

Countries that receive the most FDI are those that allow multinational corporations to set up competitive facilities able to withstand global competition. Transport costs and different preferences mean that large markets will continue to attract more investment than small ones. However, no country can afford to take for granted a continued inflow of FDI, especially high-quality, export-oriented FDI. That means that the ultimate draw for FDI is the economic base: FDI incentives and targeting cannot by themselves compensate for the lack of such a base.

Cheap unskilled labor remains a source of competitive advantage, but its importance is diminishing. Natural resources are similar. They give a competitive basis for growth as long as they are plentiful in supply and face growing demand. However, most primary exports face slow-growing markets and are vulnerable to substitution, while resource-based manufactures are among the slowest growing in world trade.

The East Asian experience, particularly that of countries like Malaysia and the Philippines, shows that FDI can be attracted to high-technology activities without any strong government strategy. What worked for them were largely good macroeconomic management, luck, and welcoming FDI policies. High-technology multinational corporations had already established a base in Singapore. The rise of the semiconductor industry and the need for cheap labor for assembling and testing the devices had led U.S. companies to seek labor overseas. Over time, Japanese and other firms joined in this quest, and soon this pattern spread to a number of other export-oriented electronics activities. Countries with low wages, stable macroeconomic regimes, good infrastructure facilities, English-speaking workers, and solid FDI incentives were able to attract investments relocating from industrial countries and from Singapore. Apart from these general attractions, therefore, FDI targeting did not play much of a role.

While multinational corporations invested in automation and skill creation in their high-technology assembly operations, sustained deepening of local content and technologies took place mainly as a result of government incentives for upgrading, supply-side support of skill and infrastructure creation, and support for local suppliers. Malaysia adopted Singapore-style strategies to induce firms to boost local content, but it did so mainly by attracting other multinational corporations rather than by upgrading a (relatively weak) local skill and industrial supplier base. There was some increase in R&D activity, but not to the levels reached by Singapore. Other countries in the region did not adopt similar proactive strategies. As a result, high-technology operations remain fairly shallow in Indonesia, the Philippines and Thailand, a constraint to their future industrial growth and competitiveness. Their governments are eager to improve their FDI targeting and upgrade local skills and supply capabilities.

POLICY INTERVENTIONS TO ATTRACT HIGHER-QUALITY FOREIGN DIRECT INVESTMENT. There is thus a case for policy interventions to attract higher-quality FDI and to induce investors to upgrade and deepen the technology content of their activities over time. The economic rationale for interventions is to lower transaction costs, compensate for deficient information, and coordinate the needs of multinational corporations, the assets of the host economy, and the potential to improve those assets.

While most FDI regimes are converging on a common (and reasonably welcoming) set of rules and incentives, there are still large differences in how these rules are implemented. The FDI approval process can take several times longer and cost several times more in one country than in another with similar policies. After approval, the cost of setting up facilities, operating them, importing and exporting goods, paying taxes, hiring and firing workers, and generally dealing with the authorities can differ enormously.

Such transaction costs can significantly affect the competitive position of a host economy. An important part of the competitiveness strategy thus consists of reducing distorting and wasteful business costs. These costs affect both local and foreign enterprises. However, foreign investors have a much wider set of options before them and are able to compare transaction costs in different countries. Thus, attracting multinational corporations requires benchmarking transaction costs against those of competing host countries. Many countries have set up one-stop promotion agencies to assist international investors in getting necessary approvals. But unless the agencies have the authority to negotiate the regulatory system, and unless the rules themselves are simplified, this may not help. Indeed, there is a risk that a "one-stop shop" becomes just "one more stop."

Despite their size and international exposure, multinational corporations face market failures in information. They collect much information on potential sites, but their information base is far from perfect and the decisionmaking can be subjective and biased. Taking economic fundamentals as given, it can be worthwhile to invest in altering the perceptions of potential investors by providing better information and improving a country's international image. The experience of Singapore (Box 2.6) suggests that promotion can be extremely effective in raising the inflow of investment and its quality.

Such promotion efforts are highly skill intensive and potentially expensive, however, and need to be carefully mounted. Targeting can be general (countries with which there are trade or historic connections, or which lack past connections but are ripe for establishing them), industry specific (industries in which the host economy has an actual or potential competitive edge), and even investor specific. Incentives play a minor role in a good promotion program—good long-term investors are not particularly susceptible to short-term inducements.

Effective promotion should go beyond simply "marketing a country" to coordinating the supply of immobile assets with the specific needs of targeted investors. Such coordination addresses potential failures in markets and institutions for skills, technical services, or infrastructure relative to the specific needs of new activities targeted through FDI. A developing country may not be able to meet such needs, particularly in activities with advanced skill and technology requirements. Attracting FDI in such industries can be facilitated if the host government identifies and meets the needs of potential investors. Singapore even involves managers in designing training and infrastructure programs, ensuring that the country remains attractive for their future high-technology investments. The information and skill needs of such coordination and targeting exceed those of promotion, requiring detailed knowledge of technologies (their skill, logistical, infrastructural, supply, and institutional needs) as well as of the strategies of multinational corporations.

Concluding Remarks

East Asia's experience with FDI and competitiveness offers rich policy lessons. FDI has enormous potential for contributing to competitiveness in host economies, particularly in technology-intensive activities and those being knit into international systems of production and distribution. With continuing technical change and globalization, the role of FDI is set to grow.

Highly interventionist policies on FDI are no longer feasible, nor are they economically desirable in the emerging technology setting—the costs and risks are too large, and countries can easily get left behind in the fastest growing activities. With most countries adopting liberal investment policies, however, a strong role remains for governments in extracting the most benefits for competitiveness from FDI. Passively opening up to multinational corporations may be useful in realizing existing, largely static, com-

Box 2.6 Managing Foreign Direct Investment Strategy in Singapore

Singapore's Economic Development Board (EDB) has centralized responsibility for FDI targeting, within the strategic direction provided by its parent organization, the Ministry of Trade and Industry. The EDB has the authority to coordinate all activities relating to industrial competitiveness and FDI and the resources to hire professional staff (essential to manage discretionary policy efficiently and honestly). Over time the agency has become the global benchmark for FDI promotion and approval procedures. Its ability to coordinate the needs of foreign investors with measures to boost local skills and capabilities has also been critical. The government conducts periodic strategic and competitiveness studies to chart the economy's industrial evolution and upgrading. Multilateral corporations are actively involved in strategy formulation and are given a strong stake in the economy's development.

Since drawing up its 1991 Strategic Economic Plan, the government has focused on *industrial clusters*, a reference to interlinked activities in a value chain, the network of vertically and horizontally linked supporting industries and resources that collectively make products or services competitive. In the manufacturing sector the government analyzes the strengths and weaknesses of leading industrial clusters and seeks to upgrade local capabilities and institutions across the entire value chain to promote the clusters' competitiveness. The program explicitly seeks to avoid the kind of industrial hollowing out Hong Kong (China) and many other industrialized economies experienced.

This strategy has allowed Singapore to become, for example, the leading production center for hard disk drives in the world, with considerable local linkages with advanced suppliers and R&D institutions. In 1994 the government set up a S$1 billion Cluster Development Fund to support specific clusters such as a new wafer fabrication park. It also launched a Co-Investment Program to provide official equity financing for joint ventures and strategic ventures, not just in Singapore but also overseas when that serves Singapore's competitive interests. The EDB can take equity stakes to support cluster development by addressing critical gaps and improving local enterprises. The government also offers start-up grants to attract multinational corporations to activities thought to be critical to particular industry clusters. Development of the hard disk drive industry, for example, led to the growth of sophisticated local suppliers such as Advanced Systems Automation (advanced wafer packaging equipment) and Manufacturing Integration Technology (semiconductor testing equipment).

parative advantages, but it will not lead to dynamic and sustained upgrading unless policies address the needs of skill, technology, and institutional upgrading. To generalize from the available evidence:

- On *liberalization*, the new rules of the game and the nature of technical change mean that globalization is practically irreversible. Countries cannot shut their economies off from participation in international trade and investment. Nor should they want to: the way to raise productivity and living standards is through greater participation—but with preparation to ensure that liberalization does not lead to economic damage or technological stagnation. The pace of liberalization has to be calibrated to ensure that domestic capabilities improve and that the productive structure upgrades. The rules of the game may provide sufficient flexibility to developing countries to manage this calibration, but careful preparation and market-friendly strategies are needed to take advantage of the opportunities.
- The need is not just to provide an attractive, low-cost, and friendly regime for foreign investors, but also *to use investment promotion efficiently and economically to promote upgrading and deepening*. Export competitiveness depends vitally on effective investment promotion institutions. There are market failures in the FDI location process. Providing better information and properly marketing a country's location advantages can make a significant difference. The skillful use of incentives and investment in skills and supply capabilities can yield large dividends in building a dynamic export structure and can induce rapid upgrading of affiliate activities. Singapore's targeting and upgrading strategies, with the participation and consent of multilateral corporations, shows how good strategy can make an enormous difference to export performance. The bottom line here is the quality, honesty, and autonomy of the investment promotion agency. Most countries have such agencies. Few have the private sector orientation, remuneration levels, management skills, and information to perform effectively.
- *Boosting domestic skills and capabilities* is probably the most important tool of competitiveness policy. Reliance on multinational corporations does not do away with the need to invest in domestic capabilities. The entry of multinational corporations can at most reduce the need to develop skills and technical capabilities in the initial stages of building competitiveness. Once this stage is complete, and industry has to tackle more demanding activities, FDI cannot substitute for local investment in capabilities.

In some cases the outcome of FDI depends significantly on how well the host economy bargains with international investors. Often, however, multinational corporations tend to have better negotiating skills and information. And many developing countries need the assets of multinational corporations more than multinational corporations need the locational advantages offered by a specific country. Particularly in export-oriented investment projects, where natural resources are not a prime consideration, multinational corporations tend to have several alternative locations. Host countries may also have alternative foreign investors, but are often unaware of this.

Thus where the outcome of an FDI project depends on astute bargaining, developing countries do poorly compared with multinational corporations. The risk is particularly large for lumpy resource extraction projects and privatizations of large utilities and industrial companies. Manufacturing projects also involve intense and prolonged bargaining, with incentives negotiated on a case-by-case basis. Although the trend is toward nondiscretionary incentives, there is still considerable scope for bargaining.

The need for regulation is also growing, especially in competition law and environmental policy. With globalization and liberalization, few tools are left to ensure competitive conduct by foreign and local firms: effective competition policy is essential (see Chapter 3). However, many developing countries are unaware of the need for competition policy. Setting such policy is a complex task, requiring specialized skills and expertise that are often scarce in developing countries.

Managing FDI policy effectively is a demanding task. While a passive approach is unlikely to be sufficient because of institutional and market deficiencies, an active approach faces difficulties in design, implementation, and monitoring. If analytic and administrative capabilities are not adequate to the demands of the task, it may be best to minimize interventions and simply reduce obstacles in the way of FDI and leave resource allocation to the market. A laissez-faire FDI strategy may yield significant benefits, particularly in a host country whose past policies have led to underperformance in competitiveness and investment attraction. A strong signal to the investment community that the economy is now open for business can attract FDI into existing areas of comparative advantage.

Being more receptive to FDI may not be enough, however. If these areas are limited or if poor infrastructure or noneconomic risk undercuts their appeal, few foreign investors will respond. In any case the benefits of any FDI are likely to be static and will run out when existing advantages are used up. To ensure that FDI is sustained and enters new activities, policies are needed to target investors and to improve the quality of local factors and facilities. Policies have to be clearly aimed at competitiveness and at removing any vestiges of traditional patterns of heavy inward-orientation and market-unfriendly measures. If capabilities are weak, they should be built up rather than abandoning strategy altogether.

There is no ideal, universal FDI strategy. An FDI strategy has to suit the conditions of the country and evolve as the country's needs and competitive position in the world change. Effective strategy requires vision, coherence, and coordination. It also requires the ability to decide on tradeoffs between different objectives of development. The strategy-making body should be located near the head of government, to give it a broad view of national needs and priorities—and the power to implement appropriate policies.

Appendix 2.1 Statistical Analysis of Strategies for High-Technology Competitiveness

This appendix presents some results of a statistical analysis of the strategic determinants of export performance in high-technology products for 73 industrial and developing countries in 1985 and 1995. [2] Although the analysis used several variables to analyze strategies, only the results for the two related to "autonomous" and "FDI-dependent" strategies are shown here.[3] The dependent variable is the revealed comparative advantage (RCA) of each country in high-technology exports (its share of world high-technology exports divided by its total share of world exports). The explanatory variables are reliance on FDI (measured as a share of domestic investment) and enterprise-financed R&D as a percentage of GNP. The analysis shows how different sets of countries achieved competitiveness in high-technology exports by investing in domestic R&D and relying on FDI.

A K-means cluster analysis is used, with countries grouped by their similarity on the three selected variables (high-technology revealed comparative advantage, R&D, and FDI).[4] The group with the highest revealed comparative advantage in high-technology exports—China, Hong Kong (China), Malaysia, Mexico, the Philippines, Singapore and Thailand, at 1.63—shows a strong dependence on FDI and a low propensity to conduct local R&D (see Figure 2.2). The second group—Korea and Taiwan (China), at 1.47— shows moderate FDI inflow but very strong R&D. The OECD group (0.86) shows a balance between the two. This analysis illustrates that the FDI-dependent strategy has led to a stronger performance in high-technology manufactured exports.

Figure 2.3 shows changes in revealed comparative advantage in high-technology exports, R&D, and FDI propensities over 1985–95 for countries with better-than-average revealed comparative advantage. The clusters now turn out to be slightly different. In 1995 the cluster comprising Hong Kong (China), Malaysia, Singapore, and Taiwan (China) now has the highest revealed comparative advantage (1.82). This group started with a very high average reliance on FDI in 1985 and increased it further while also raising its R&D intensity. Next is Korea, which does not cluster with any

Figure 2.2 Revealed Comparative Advantage in High-Technology Exports and Main Industrial Strategies, 1995

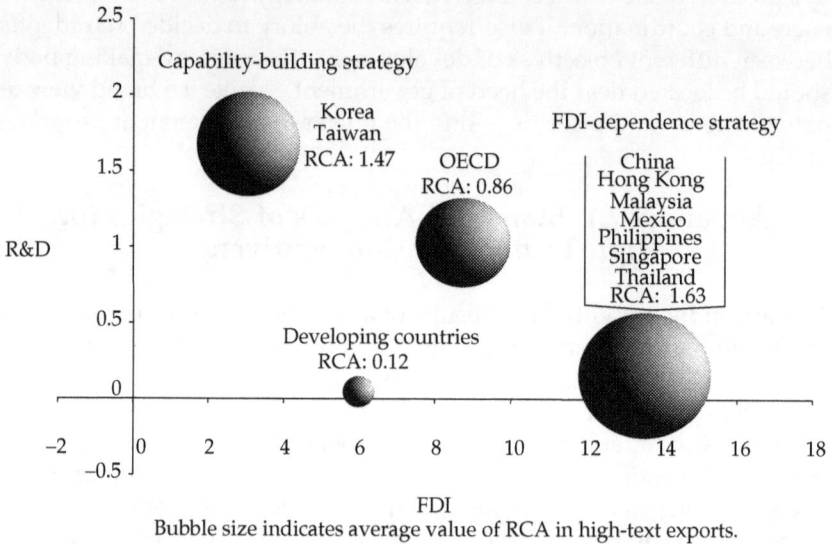

Bubble size indicates average value of RCA in high-text exports.

Source: Author's computation.

other country. Its RCA rises rapidly with its R&D intensity, while its reliance on FDI remains almost constant. Following is the group with China, Hungary, Mexico, the Philippines, and Thailand, with a moderately high and rapidly rising RCA, very low and stagnant R&D propensity, and sharply increasing reliance on FDI. Of the OECD countries, Germany, Japan, and Switzerland cluster together, with a positive but slightly diminishing RCA, high and rising R&D, and low and stable reliance on FDI. The other OECD countries have a stable RCA of below unity and increasing R&D and FDI propensities; however, FDI rises much faster than R&D.

These data suggest that strategic differences between countries that rely on FDI and those that invest in local R&D persist over time (and apply to industrial as well as developing economies). Policy liberalization is leading to similar FDI regimes across the world, but traditional differences continue to matter.[5] *OECD Science, Technology and Industry Scoreboard 1999* (OECD 1999) remarks on the large differences between FDI shares in production, R&D, and exports within high-income countries, attributable partly to size and location, but also to path dependence in business attitudes and

Figure 2.3 Changes in the Revealed Comparative Advantage of High-Technology Exports and Industrial Strategy Paths for Countries with Above-World Average Revealed Comparative Advantage, 1985–95

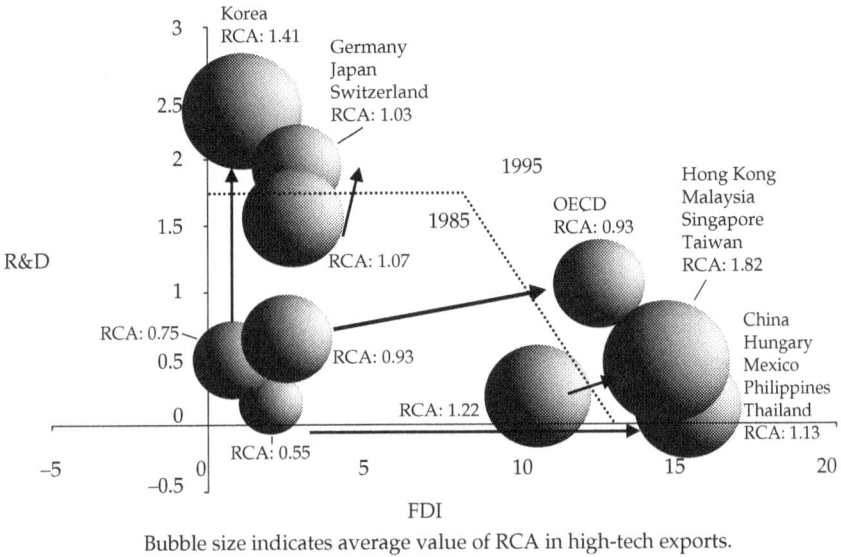

Bubble size indicates average value of RCA in high-tech exports.

Note: OECD data exclude Mexico and the Republic of Korea.
Source: Author's computation.

policy. It may nevertheless be that the current wave of globalization and technical change, with cross-border mergers and acquisitions spearheading the spread of international production systems, will knit economies much closer together.

Notes

1. The ASEAN 10 are Brunei, Cambodia, Indonesia, Lao PDR, Malaysia, Myanmar, the Philippines, Singapore, Thailand, and Vietnam.

2. I am grateful to Manuel Albaladejo for providing this analysis.

3. The variables included technical skills (numbers enrolled in tertiary technical subjects as a percentage of the population), enterprise-financed R&D as a percentage of gross national product, FDI as a percentage of gross domestic investment for the relevant period, and manufacturing wages and gross domestic

investment as a percentage of gross domestic product. The data on export performance by technology levels are analyzed in detail in Lall (2000).

4. K-means cluster analysis is used to identify relatively homogeneous groups, in this case groups of countries with similar export patterns (dependent variable) and strategies (independent variables). Variables values are not weighted to maintain impartiality. The number of groups was specified in advance at five clusters. Squared Euclidean distance, the sum of the squared differences over all of the variables, was used to identify five initial cluster centers as a reference point for the other cases. Since good cluster centers separate the cases well, the strategy was to chose five cases (groups of countries) with large distances between them, using their squared Euclidean values as the initial cluster centers. Other cases then group to the cluster with the smallest distance between the case and the cluster center. The algorithm used for determining clustering membership is based on nearest centroid sorting. Once the cases are classified, the final cluster centers give the average values of the variables for countries in the clusters.

5. The data end at 1995 and so do not capture the burst of FDI in countries like Korea in later years.

References

Lall, Sanjaya. 2000. "The Technological Structure and Performance of Developing Country Manufactured Exports 1985-98." *Oxford Development Studies* 28(3): 337–69.

OECD (Organisation for Economic Co-operation and Development). 1999. *OECD Science, Technology and Industry Scoreboard 1999: Benchmarking Knowledge-Based Economies.* Paris.

UNCTAD (United Nations Conference on Trade and Development). 2000. *World Investment Report 2000: Cross-Border Mergers and Acquisitions and Development.* New York: United Nations.

3

Corporate Governance, Corporate Performance, and Investor Confidence in East Asia

Ijaz Nabi and Behdad Nowroozi

The East Asian crisis was in part a crisis of the corporate sector. Corporations had grown rapidly as external financing became readily available, and most firms resorted to excessive debt financing to meet their often imprudent investment needs. With burgeoning debt-equity ratios (the average debt-equity ratio for Thai firms, for example, doubled from 1.5 in 1994 to 3 in 1997), the highly leveraged corporations also became less efficient (Table 3.1). In 1992–96 assets grew rapidly, while economic value added declined. Profitability failed to keep pace with debt-financed growth, and most corporations' ability to service their debt declined, as reflected in declining interest cover, By 1998 the percentage of corporations in distress had grown alarmingly.

Table 3.1 Corporate Performance in Four East Asian Countries before the Crisis, 1992–96
(*percent*)

Country	Average change in fixed assets	Economic value added in the period[a]	Decrease in interest cover[b]	Percentage of corporations in distress in 1998
Indonesia	33	–9	—	45.6
Korea, Rep. of	17	–2	1.42–1.07	31.5
Malaysia	26	3	9.1–6.7	—
Thailand	29	–8	4.6–1.9	—

— Not available.
a. Pretax return on assets less the cost of capital.
b. Ratio of operating profit to interest payable.
Source: Pomerleano, 1998.

Would corporate performance have been different had corporate expansion been mainly equity financed rather than debt financed? Empirical evidence suggests the answer is yes. Firms that raise external financing successfully tend to grow faster, while success in raising capital in the stock market is associated with higher productivity (Demirgüç-Kunt and Maksimovic 1994).

East Asian firms resorted to heavy debt finance because capital markets were underdeveloped, in part because the institutional and regulatory frameworks for external nondebt finance were underdeveloped and in part because corporate governance was weak. Minority shareholder rights were ill-defined; directors and managers were not held accountable; and poor accounting, auditing, and disclosure standards and practices meant little transparency in company balance sheets. Also, large blockholders—usually members of one family—played a dominant role in corporate decisionmaking. These weaknesses discouraged prospective investors and were detrimental to the development of capital markets. Shareholders that did take equity positions in firms under these circumstances found it difficult to enforce discipline on majority owners to prevent the poor investment decisions that led to the decline of profitability evidenced in Table 3.1.

Structural Influences on Corporate Performance

Concentration of ownership is commonly thought to facilitate corporate decisionmaking and to reduce inefficiency and abuse. Studies have also shown that concentration can lead to problems. When ownership is concentrated, market discipline on large shareholders is attenuated, often leading to empire building, undue risk-taking, and excessive diversification, especially in developing countries, where the regulatory environment is ineffective and disclosure practices and property rights are weak. Concentration of ownership also promotes uncomfortably close relations with financial institutions, which can lead to risky investments. Table 3.2 shows the percentage of ownership concentration in selected countries in Asia and Latin America.

A recent study (Alba, Claessens, and Djankov 1998) of the relationship between ownership concentration and profitability among Thai firms during 1992–96 found ownership concentration to be positively related to profitability in 1992 but negatively related by 1996 (Figure 3.1). The negative correlation results from the higher leverage (higher debt-equity ratios) of firms with concentrated ownership, which is negatively correlated with profitability. Since the sample of firms is the same for both periods, the results suggest a deteriorating performance for firms with concentrated ownership relative to firms with less concentrated ownership.

Table 3.2 Ownership Concentration in the 10 Largest Firms for Selected Regions and Countries
(*percent*)

Country	All firms	Largest 10 firms	Country	All firms	Largest 10 firms
Asia			*Latin America*		
India	38	40	Argentina	50	53
Indonesia	53	58	Brazil	31	57
Korea, Rep. of	23	20	Chile	41	45
Malaysia	46	20	Colombia	63	63
Pakistan	26	54	Mexico	64	64
Philippines	56	37	Venezuela	—	51
Sri Lanka	60	60			
Thailand	44	47			

— Not available.

Note: Ownership concentration is the average percentage of common shares owned by the three largest private shareholders in the 10 largest nonfinancial firms. The percentages are not corrected for shareholder affiliation and cross-shareholding between firms.

Source: La Porta and others, 1998.

Figure 3.1 Relationship of Ownership Concentration and Profitability, 1992 and 1996

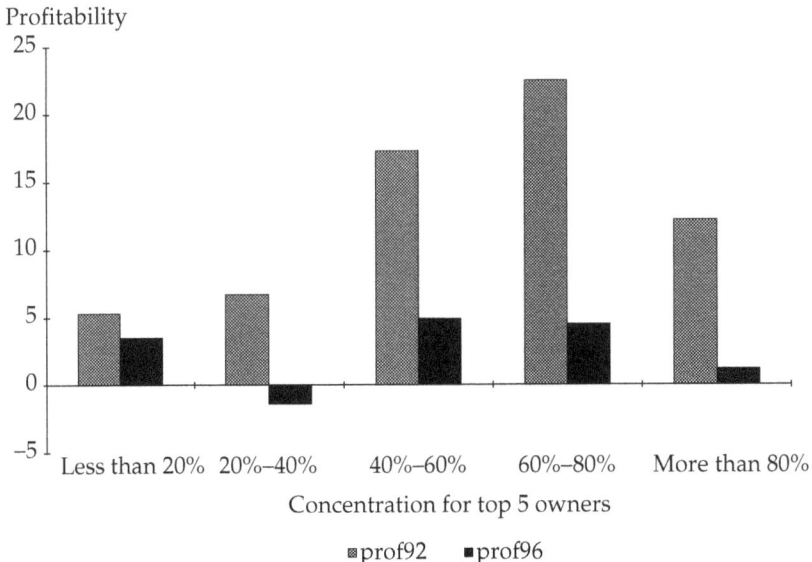

Concentration for top 5 owners

Source: Alba, Claessens, and Djankov, 1998.

There is broad agreement that the rapid expansion of East Asian corporations led to diversification into unrelated activities in which they had little expertise. Such diversified conglomerates with concentrated ownership tend to be inefficient. Their shares trade at a discount, and they have lower book value than independent firms in related industries. When such conglomerates are broken up, share prices often rise (Rajan and Zingales 1997). Inefficiencies arise for a variety of reasons. Majority owners and managers like to build empires, so internal management suffers. And resources get diverted to inefficient units. The results are a decline in profitability.

Such actions are possible because of weak corporate governance. Poor information about management decisions and weak protection of minority shareholder rights result in inadequate surveillance of management decisions and their impact on corporate balance sheets. Such forbearance eventually leads to a crisis of confidence, often followed by the wholesale recall of loans, as happened in East Asia in 1997–98.

Benchmarking Corporate Governance

Improving corporate governance thus depends on protection of minority shareholder rights, protection of creditor rights, and sound accounting standards that ensure transparency and disclosure.

Protection of Minority Shareholders

Only if shareholders have clearly defined and enforceable rights can they impose financial and economic discipline on corporate insiders. This requires sensible institutional procedures and a supportive legal and judicial system. On several indicators of investor and creditor protection developed by La Porta and others (1997, 1998), Malaysia scores consistently higher than three other East Asian countries, Indonesia, the Philippines, and Thailand (Table 3.3). Those three countries also score lower than the average for Asia on most of these indicators.

Protection of Creditors

When corporations rely heavily on debt finance, creditors need adequate legal rights to enforce discipline on borrowers. The objective should not be large-scale, court-supervised bankruptcy, but rather the credible threat of such action so that borrowers have an incentive to behave responsibly. Table 3.3 compares Asian and Latin American countries on several attributes of creditor protection and judicial enforcement; Table 3.4 compares how well the judicial system protects creditors in a sample of East Asian

Table 3.3 Indexes of Investor and Creditor Protection in Asia and Latin America

Country	Investor pro-tection[a]	Creditor pro-tection[b]	Judicial enforce-ment[c]	Country	Investor pro-tection[a]	Creditor pro-tection[b]	Judicial enforce-ment[c]
Asia				*Latin America*			
India	2	4	6.1	Argentina	4	1	5.6
Indonesia	2	4	4.4	Brazil	4	1	6.5
Malaysia	4	4	7.7	Chile	4	2	6.8
Pakistan	5	4	4.3	Colombia	1	0	5.7
Philippines	4	0	4.1	Mexico	0	0	6.0
Sri Lanka	2	3	5.0	Venezuela	1	—	6.2
Thailand	3	3	5.9	Average	2.2	0.8	6.1
Average	3.1	3.1	5.4				

— Not available.

a. An index of how well the legal framework protects equity investors. A score of 6 means that shareholders may vote by mail and are not required to deposit shares in advance of a meeting, cumulative voting is allowed, less than 10 percent of share capital is required to call a meeting, oppressed minority protection is in place, and legislation mandates one vote per share for all shares (or equivalent).

b. An index of how well the legal framework protects secured creditors. A score of 4 means that firms face minimum restrictions (creditors' consent, for example in filing for reorganization; there is no automatic stay on collateral; debtors lose control of the firm during a reorganization; and secured creditors receive priority during a reorganization.

c. An index of the quality of judicial enforcement, ranging from 1 to 10 (best), based on five subindexes: efficiency of the judicial system, rule of law, risk of corruption, risk of expropriation, and risk of contract repudiation.

Source: La Porta and others, 1997, 1998.

economies, using a different set of indicators. Within the region, Hong Kong (China) was among the highest-scoring countries on attributes of creditor rights, while the Republic of Korea and Singapore scored highest on judicial efficacy.

Transparency and Disclosure

Transparency and disclosure are fundamental to the soundness of financial markets and to responsible decisionmaking by firms. The ability of investors and lenders to enforce market discipline on firms' majority share-

Table 3.4 Creditor Rights and Efficacy of the Judicial System in East Asia

Economy	Creditor rights[a]	Judicial efficacy[b]
Hong Kong (China)	4	5.5
Indonesia	0	3.5
Japan	3	5.5
Korea	3	6.5
Malaysia	3	4.5
Philippines	0	2.0
Singapore	4	7.0
Taiwan (China)	4	3.5
Thailand	2	5.5

a. Summary of four dummy variables. A score of 4 means that judgments are rendered in less than 90 days, incumbent management does not stay during a restructuring or bankruptcy, there is no automatic stay on assets, and secured creditors have the highest priority in payment.
b. Average of four variables each for restructuring and liquidation measuring expense, ease, efficiency, and speed, with a maximum score of 8 each for restructuring and liquidation. For example, on speed, 0 points are assigned if restructuring is very slow, 1 if slow, and 2 if quick. Similar rankings are constructed for the other three variables.
Source: Claessens and others (1999), based on data from Asian Development Bank, 1999.

holders and managers depends on the quality of information that firms disclose. Bank supervisors and regulators who implement a country's prudential rules governing banking and securities activities also need reliable information to check violations of the public trust.

Transparency and disclosure rest on reliable accounting standards and auditing practices. In response to the deepening of financial markets, Malaysia, Korea, and Thailand improved their accounting standards to bring them in line with those set by the International Accounting Standards Committee. This required the establishment of or improvements in professional institutions such as the Korean Accounting Standards Board.

Table 3.5 shows the rankings of selected industrial and East Asian economies on the eve of the crisis on an index of accounting standards. The index reflects the inclusion in company annual reports of 90 factors that accountants identify as useful indicators of a company's financial affairs. Malaysia stands out for its high ranking.

Table 3.5 A Comparison of Accounting Standards for Selected Economies, 1998

Economy	Ranking by an index of accounting standards
Australia	75
Hong Kong (China)	69
Indonesia	—
Japan	65
Malaysia	76
New Zealand	70
Philippines	65
Singapore	78
South Korea	62
Taiwan (China)	65
Thailand	64
United States	71

— Not calculated.
Note: The higher the index number, the higher the ranking.
Source: Dyck, 2000.

Reform of Corporate Governance in East Asia and the Challenges Ahead

In the aftermath of the Asian crisis in 1997, three countries have embarked on significant reforms of their corporate sectors, including corporate governance. Korea, Malaysia, and Thailand have improved corporate governance practices in varying degrees. Korea and Malaysia have made the greater progress.

Malaysian Reforms: Starting from a Sound Base

Corporate governance is reasonably strong in Malaysia, as Tables 3.3, 3.4, and 3.5 indicate. A recent World Bank report (2001) on the observance of standards and codes of corporate governance comes to a similar conclusion (Appendix 3.1) Nonetheless, the government accelerated the reform program in response to the difficulties firms experienced during the financial crisis of 1997–98 (Thillainathan 1999).

The reform strategy has focused on improving laws and regulations on corporate governance and financial reporting and disclosure, intensifying enforcement of securities laws and regulations, and strengthening institutions. A high-level Committee on Corporate Governance, consisting of members of the Securities Commission and the Kuala Lumpur Stock

Exchange, was established to recommend improvements in corporate governance laws and practices. Changes were introduced on prospectus disclosure, civil actions by investors against directors, and accreditation of directors. The Securities Commission Act was amended to require enhanced disclosure on issuers, to impose stringent sanctions for false or misleading information in prospectuses, and to allow investors to file civil suits against a company's directors and its advisers where the law has been violated.

A new voluntary Code on Corporate Governance sets out broad principles for good corporate governance practices for listed companies. Listed companies are required to disclose in their annual reports the extent of their compliance with the code.

Restructuring of the enforcement department of the Securities Commission has resulted in improved enforcement. In 1999 the commission investigated 54 cases and prosecuted 23 of them for offenses ranging from submission of false or misleading information to acts to defraud or short-sell. The Securities Commission now has the power to apply to the court for disqualification of chief executives and directors of listed companies that violate securities law or listing rules. It can also take civil action against anyone involved in insider trading or market manipulation. Amendments to the Securities Industry Act allow the Kuala Lumpur Stock Exchange to take action against directors or others involved in violating listing rules. Other measures have also been implemented to improve transparency and disclosure (Box 3.1).

Despite these reforms, however, minority shareholders are still exposed to considerable risk as long as shareholding remains concentrated and directors remain under the undue influence of owner-managers and so are not truly independent. The challenge is to see that the judiciary to which aggrieved parties turn for redress operates impartially and that laws are enforced effectively. Greater efforts are required to ensure that the judiciary works independently and impartially while maintaining a high degree of integrity. Equally important are measures to curb monopolistic behavior and concentrated shareholding. Continuing to open markets to competition is essential for reducing incentives for ownership concentration; so is increasing the incentives for dispersed shareholding, risk diversification, and improved corporate governance practices.

Reform in Thailand: Starting from a Weaker Base

There is broad recognition in Thailand that poor systems of corporate governance contributed to the financial crisis by shielding banks, specialized financial institutions, and corporations from market discipline. Inadequate disclosure, poor audit practices, ineffective boards of directors, weak internal controls, unreliable financial reporting, inadequate protection of minor-

---

Box 3.1 Progress in Improving Shareholders' Rights and Financial Reporting and Disclosure in Malaysia

- The Securities Industry Act has been amended to strengthen regulators' ability to take action against directors and others. Directors and major shareholders, and management of listed companies have statutory duty to comply with Kuala Lumpur Stock Exchange listing rules.
- The definition of insiders has been expanded. Insiders are no longer defined only as people with fiduciary or confidentiality duties (directors, mangers, advisers, agents), but include all people who possess material nonpublic information.
- Listing rules now require companies to appoint an independent corporate adviser to watch out for the interests of minority shareholders.
- A Code of Corporate Governance has been issued.
- The Malaysian Accounting Standards Board has been established as the sole authority to set accounting standards. Compliance with standards is required by law.
- Accounting standards have been improved.
- All listed companies must have audit committees composed of three members, the majority of whom must be independent.

---

ity shareholder rights, and lax enforcement of compliance are reflected in Thailand's low ranking in cross-country benchmarking on corporate governance indicators (see Tables 3.3, 3.4, and 3.5).

Since the financial crisis the government has focused on streamlining institutional arrangements, improving the reliability of financial information and disclosure, strengthening corporate board oversight and effectiveness, increasing shareholder rights, improving the effectiveness of the legal and regulatory framework, and intensifying enforcement of laws and regulations related to public companies (Nabi and Shivakumar 2001).

Amendments to the Public Companies and Securities and Exchange Acts increase shareholders' rights and the ease with which they exercise those rights, improve accountability of boards of directors and officers, and increase enforcement. Efforts are under way to strengthen the institutions responsible for ensuring accountability and compliance, including professional organizations, and to improve and enforce such mechanisms as the code of ethics, the code of conduct for accountants, and the code of best practices for directors. All listed companies are now required to have an audit committee for their board of directors.

An Institute of Directors has been established to provide training for directors and executives on their duties and responsibilities. By the end of 2001 more than 180 executives of Thai companies had completed the institute's training program for company directors. Significant work has been done on improving accounting and auditing standards and practices in the last few years (Box 3.2).

The most critical challenge in accounting and auditing reform in Thailand is implementation—changing from a culture of minimal disclosure to one of adequate disclosure. Beyond the adoption of international standards, this shift involves strengthening institutions and establishing mechanisms to ensure that the standards are applied properly.

Amendment of laws such as the Public Companies Act and the Security and Exchange Act has been very slow. So has amendment of the Accounting Professional Act to allow establishment of the Thailand Financial Accounting Standards Board and strengthening of the Institute of Certified Accountants and Auditors. Now that the corporate governance framework is being put into place, efforts should focus on transparency of busi-

---

Box 3.2 Progress on Accounting and Auditing Reform in Thailand

- Improved auditing standards consistent with international standards have been put into practice, and year 2000 accounts were audited in accordance with these standards. As of July 2001, 18 accounting standards consistent with international standards had been adopted, and 6 more were issued and expected to become effective at later dates.
- Specific rules on accounting and disclosure for banks have been drafted and are yet to be issued by the Bank of Thailand.
- The draft Accounting Professionals Act, which grants more independence to the Institute of Accountants and Auditors of Thailand (ICAAT) and allows establishment of the Thailand Financial Accounting Standards Board, was approved by the Cabinet and still needs to be approved by the Parliament.
- Audit reports, which were prepared in two paragraphs before 1998, are now prepared in three paragraphs, including a paragraph explaining management responsibility in preparing financial statements, making them consistent with international best practice.
- The Accounting Act has been amended to remove the burdensome requirements for submitting audit reports for limited partnerships, enhance disciplinary measure for accountants, and require compliance with ICAAT standards.

ness practices and promotion of business ethics. Efforts are also needed to promote good corporate governance among nonlisted companies, especially medium-size firms. Minority shareholder rights must be strengthened, and accountability and liability of directors for breach of duty must be further clarified.

Reform in Korea: Focusing on Affiliation and Cross-Ownership of Firms

Affiliation and cross-ownership among firms has characterized Korea's corporate ownership structure. Families that run *chaebols* own less than 50 percent of related companies, but they have had almost total control over the combined business group (La Porta and others 1998). Ownership concentration combined with weak protection of outsider shareholders, an uncompetitive financial system, and opportunities for malfeasance and corruption by powerful insider shareholders have led to corporate inefficiencies. In many instances, little distinction is made between managers and owners. The collusion between *chaebols* and the government has further exacerbated failures of corporate governance. Large shareholders have been reluctant to install mechanisms that could erode their control and the benefits it brings.

Since the crisis, Korea has made significant progress on transparency and financial accountability (Box 3.3). Accounting and auditing standards have been improved, and the Korean Accounting Standards Board has been established. Large listed companies have been required to establish audit committees of the boards of directors and to produce quarterly reports. A Code of Best Practices for Corporate Governance has been issued, and cross-guarantees within the *chaebol* have been eliminated.

Significant improvements have also been made in protecting shareholder rights, such as reducing thresholds for exercising shareholder rights, introducing outsider directors, and clarifying duties (Box 3.4). Protection of creditors' rights has also improved. Major reforms include establishment of the Specialized Bankruptcy Court and a Bankruptcy Commission to assist in case management. The processing period for bankruptcy cases has been shortened.

While these reforms have been critical, the reform agenda in Korea remains incomplete. The challenge now is to change corporate culture and practices through implementation and effective enforcement of the legislative changes introduced in the last few years. Going beyond changes in legislation affecting corporate governance, the government should take additional steps to promote transparency of business practices, increase and enforce shareholder rights, and improve monitoring of related-party transactions.

Box 3.3 Progress on Accounting and Auditing Reform in Korea

- The Korean Accounting Standards Board was established in September 1999 as an independent body responsible for setting accounting standards.
- The Korean Institute of Certified Public Accountants was strengthened as a professional body, and relevant laws and regulations were amended.
- Improved accounting and auditing standards consistent with international standards were issued.
- Preparation and audit of financial statements in accordance with improved standards were required beginning in 1999.
- Specific accounting standards were issued for commercial banks.
- Laws and regulations were amended, and beginning in 2000 banks and large listed companies were required to establish audit committees of the boards of directors.
- Enforcement of enhanced disciplinary measures for accountants are under way.

Box 3.4 Progress on Improving Shareholders' Rights in Korea

- Regulations on securities listings were amended to require that a quarter of boards of directors of listed companies be outside directors. (This requirement will be increased to 50 percent by 2002.)
- Listed companies must file semi-annual and quarterly reports in addition to the annual reports previously required.
- Shareholders have the right to propose matters for consideration at general meetings of shareholders.
- Cumulative voting is permitted, unless explicitly revoked in a company's articles of incorporation.
- Levels of shareholdings required for asserting certain rights have been reduced. These rights include demanding removal of directors and auditors, seeking an injunction against a director, initiating a derivative lawsuit, convening a general meeting of shareholders, inspecting account books, applying to court for appointment of an inspector to investigate company affairs, and demanding removal of a liquidator.
- Explicit fiduciary duties of directors now include performing duties for the good of the company. "Shadow directors" are subject to the same duties and liabilities as directors.
- Institutional investors are permitted to exercise voting rights directly.
- A code of Best Practices for Corporate Governance was issued.
- Corporate governance of banks and financial institutions was strengthened.

Concluding Remarks

There is a need to persist with reform. A recent investor survey of six Asian economies—Indonesia, Japan, Korea, Malaysia, Taiwan (China), and Thailand—by McKinsey & Company (1999) underscores the need to carry on with reform. Both Asian and U.S.-based investors asserted that much remains to be done in the region to catch up with the best practices (Table 3.6).

The McKinsey survey shows that institutional investors are prepared to pay a premium for good corporate governance. Institutional investors would be willing to pay a markup of more than 20 percent for shares of companies in Asia that demonstrate good corporate governance, from 20.2 percent in Japan and Taiwan (China) to as high as 27.1 percent in Indonesia. The premium is 18.3 percent in the United States and 17.9 percent in the United Kingdom. Asian corporations have much to gain from strengthening investor confidence by making the transition to sound corporate governance.

Table 3.6 Investor Opinion Surveys on Corporate Governance, 1999

Economy	Perceived quality of corporate governance from 1 (low) to 5 (high) by		Premium for good corporate governance
	Asian investors	U.S. investors	
Japan	2.9	2.2	20.2
Indonesia	1.1	1.1	27.1
Korea, Rep. of	2.2	1.8	24.2
Malaysia	—	—	24.9
Taiwan (China)	2.6	2.3	20.2
Thailand	1.8	1.3	25.7
United Kingdom	—	—	17.9
United States	—	—	18.3

— Not available.
Note: Rated on a scale from 1, very poor, to 5, very good.
Source: McKinsey & Company, 1999.

Appendix 3.1 Summary Fact Sheet on Corporate Governance in Malaysia

Aspect of corporate governance		*Remarks*
Market and Regulatory Overview		
Market capitalization (percent of GDP)		As of 12/30/99: 553 billion ringgit ($145.4 billion), or approximately 184% of 1999 GDP.
Turnover ratio		34 percent.
Number of listed companies		757 as of 12/30/1999.
Legal system (origin)		Common law system with a comprehensive legal framework.
Autonomy of capital markets regulator		Securities Commission chair and members appointed by and report to the minister of finance. Accounts annually tabled in parliament. Self-funded.
Powers of the capital markets regulator		Administrative, including powers to conduct investigations and to prosecute with consent of attorney general. No judicial powers.
Stock exchange governance		Kuala Lumpur Stock Exchange, four board members chosen by minister of finance and five by its members.
Corporate ownership structure		Supervised by the Securities Commission. Concentrated. In half the listed companies the five largest shareholders own more than 60 percent of shares.
Shareholders' Rights		
Voting rights		Each ordinary share carries one vote. Nonvoting preferred shares are rare. Some companies have "special shares," requiring holder's consent for some matters or conferring rights over board appointments.
Proxy voting	Yes	No need for notarization. Deposited before meeting; no postal ballot.
Cumulative vote, proportional representation	No	Voluntary Code on Corporate Governance sets out best practice on proportional representation.

Appendix 3.1 (continued)

Aspect of corporate governance		Remarks
Ownership shares required to call shareholders meeting	Yes	Two or more members holding 10 percent or more may call extraordinary meeting.
Redress against violations; minority oppression remedies	Yes	Personal actions, representative actions and derivative actions. Procedural difficulties with recovery of damages in representative action.
Takeover code Mandatory tender offer in change of control	Yes	Required at more than 33 percent of share capital. The bidder must pay the highest price paid for the shares of the offeree in the preceding six months.
Insider trading and self-dealing prohibition	Yes	"Insider" includes all persons with material nonpublic information. Investors allowed to seek full compensation. One case pending in court.
Preemptive rights		Kuala Lumpur Stock Exchange listing requires preemptive rights to be worked into the articles of association.
Oversight of Management		
Board structure		One tier board, combination of executive and nonexecutive.
Independent directors	Yes	1998 survey indicated most companies have good mix. Anecdotal evidence suggests controlling shareholders sometimes act as "shadow directors," in which case the Companies Act imposes fiduciary duties.
Committee practices	Yes	Getting proof and enforcing the law are difficult in practice. Audit committee, with at least three members; majority of independent directors.
Disclosure and Transparency		
External auditors	Yes	Appointed/removed at annual shareholders meeting.

(*Appendix continues on the following page.*)

Appendix 3.1 (continued)

Aspect of corporate governance		Remarks
Consolidated statements	Yes	Required by listing requirements of Kuala Lumpur Stock Exchange.
Segment reporting	Yes	Compliance is a statutory requirement.
Disclosure of price-sensitive information	Yes	Must disclose material information to the public immediately, clarify and confirm rumors and reports, and respond to unusual market action.
Accounting standards and enforcement	Yes	Compliance with International Accounting Standards in all material aspects. Compliance if a statutory requirement.
Company officers related disclosures	Yes	Aggregate remuneration is disclosed.
Related-party transactions	Yes	International. Accounting Standard (IAS24) on related party disclosures adopted; there are rules for both directors and related parties.
Disclosure of ownership		Lowered from 5 percent to 2 percent, but ownership structure is difficult to capture.
Risk management and other disclosures		Soon to be introduced. Kuala Lumpur Stock Exchange listing requirements will include annual reporting on the state of internal controls in a company.

Source: World Bank, 2001.

References

Alba, Pedro, Stijn Claessens, and Simeon Djankov. 1998. "Thailand's Corporate Financing and Governance Structures: Impact on Firm's Competitiveness." Policy Research Working Paper 2003. World Bank, Economic Policy Unit, Finance, Private Sector, and Infrastructure Network, Washington, D.C.

Asian Development Bank. 1999. *Insolvency Law Reform*. Manila.

Claessens, Stijn, Simeon Djankov, Joseph P. H. Fan, Larry H. P. Lang. 1999. "Corporate Diversification in East Asia: The Role of Ultimate Ownership and Group Affiliation." Policy Research Working Paper 2089. World Bank, Financial Economics Unit, Financial Operations Vice-Presidency, Washington, D.C.

Demirgüç-Kunt, Asli, and Vojislav Maksimovic. 1994. "Capital Structures in Developing Countries: Evidence from Ten Countries." Policy Research Working Paper 1320. World Bank, Economic Policy Unit, Finance, Private Sector, and Infrastructure Network, Washington, D.C.

Dyck, I. J. Alexander. 2000. "Ownership Structure, Legal Protections, and Corporate Governance." In Boris Pleskovic and Joseph E. Stiglitz, eds., *Annual World Bank Conference on Development Economics 2000*. Washington, D.C.: World Bank.

La Porta, Rafael, Florencio López-de-Silanes, Andrei Shleifer, and Robert Vishny. 1997. "Legal Determinants of External Finance." *Journal of Finance* 52: 1131–50.

———. 1998. "Law and Finance." *Journal of Political Economy* 106: 1113–55.

McKinsey & Company. 1999. "International Investor Opinion Survey." In cooperation with Institutional Investor's Asia Pacific Institute, U.S. Institute, and European Institute. London.

Nabi, Ijaz, and Jayasnakar Shivakumar. 2001. "Back from the Brink: Thailand's Response to the 1997 Economic Crisis." World Bank, East Asia and Pacific, Poverty Reduction and Economic Management Sector Unit, Washington, D.C. Processed.

Pomerleano, Michael. 1998. "The East Asia Crisis and Corporate Finances: The Untold Micro-Story." Policy Research Working Paper 1990. World Bank, Development Prospects Group, Washington, D.C.

Rajan, Raghuram, and Luigi Zingales 1997. "Power in a Theory of the Firm." NBER Working Paper 6274. Cambridge, Mass.: National Bureau of Economic Research.

Thillainathan, R. 1999. "Corporate Governance and Restructuring in Malaysia." Paper prepared for the joint World Bank–Organization for Economic

Cooperation and Development survey of corporate governance arrangements in selected Asian countries. Washington, D.C., and Paris. Processed.

World Bank. 2001. "Reports on the Observance of Standards and Codes (ROSC)." Corporate Governance Modules. World Bank, Corporate Governance Unit, Private Sector Advisory Services Department, Washington, D.C. Processed.

4
Competition Policy, Economic Adjustment, and Competitiveness

R. S. Khemani

The economic and financial crises of the late 1990s in several developing and transition market economies underscored the importance of a flexible, dynamic, and competitive private sector for fostering sustained and widely shared economic development. Just as good public governance (public administration, taxation, and economic regulation) is important for the efficient and accountable delivery of government-provided services, so is a competitive business environment for private sector firms important for economic development. Competition policy has a critical role to play in fostering such an environment. By maintaining and encouraging competition, it results in more efficient allocation of resources and greater competitiveness.

Competition policy encompasses government measures that directly affect the behavior of enterprises and the structure of industry. These measures include policies to enhance competition in local and national markets (such as liberalized trade policy, relaxed foreign investment and ownership requirements, and economic deregulation) as well as competition law (also referred to as antitrust or antimonopoly law) to prevent anticompetitive business practices by firms and unnecessary government intervention in the marketplace (Khemani and Dutz 1996). Some 82 countries have competition laws on the books. More than half have adopted or strengthened such policies since the early 1990s, but resource constraints and administrative capacity can affect implementation.

Effective competition forces firms to focus on efficiency and offer a greater choice of higher-quality products and services at lower prices. It also diminishes distorted price and profit signals and the risks of misguided investment and output decisions, which can have economywide repercussions—as the economic crises in Indonesia, the Republic of Korea, Russia, and Thailand (among others) have demonstrated. Competitive markets offer much more information on business opportunities and on decisions made by firms. They therefore encourage the movement of resources from lower- to higher-value uses; promote greater accountability and transparency in government-business relations and decisionmaking; and reduce corruption, lobbying, and rent-seeking behavior.

Additionally, competitive markets provide opportunities for broad-based participation in the economy and for sharing in the benefits of economic growth. Without competition some firms are more likely to possess market power that enables them to earn excess profits and wield political influence to tilt public policy in their favor. Such a business environment is inimical to a culture of good corporate governance in which corporate management acts to maximize the firm's value and is accountable to its investors and other stakeholders (see Chapter 3 for a discussion of corporate governance in East Asia).[1] As a result, businesses find it difficult to build investor confidence, attract domestic and foreign investment, and develop a capital market and a cadre of professional managers. Ultimately, these shortcomings can undermine the competitiveness of an economy and lead to social and political instability.

Developing and transition market economies need to take credible measures to make competition policy an integral part of the government's policy framework to foster competitiveness and economic development. This chapter identifies structural and governance features common in many developing and transition market economies and discusses their implications for economic development in previously high-performing East Asian countries as well as in other developing and transition market economies. The chapter also briefly describes best practice in modern competition law policy.

Structural and Governance Features of Developing and Transition Market Economies

Until the late 1990s several developing and transition market economies had an impressive record of fostering investment, exports, economic growth, and poverty reduction. In East Asia these economies were called "tiger" economies. Questions were frequently raised in the economic literature about the superiority of the "East Asian economic" model over the "Anglo-Saxon capitalist market" model for fostering development. Many of the East Asian nations did not have the fiscal deficits, major current account or trade imbalances, high inflation rates, or excessive wage settlements common at the time in countries following the Anglo-Saxon model.

Why the Faltering Performance?

The East Asian development model yielded impressive results. With a reliance on exports, good infrastructure, and the high education and skill levels of the labor force, these economies were viewed as relatively outward oriented and internationally competitive. But with the exception of Korea and Thailand, most of them lacked formal competition laws intended to foster domestic competition by preventing restrictive business practices and anticompetitive market structures. And even in Korea and Thailand, effective application of competition law was lacking.

This reliance on an imperfect competition model for both domestic and international (export) market conduct and the inherent inefficiencies in such a system eventually undermined the economic progress achieved. In Latin America and the former centrally planned economies in Europe, structural reforms yielded some initial positive results, but the development of competitive market-oriented economies was impeded by inadequate legal and economic frameworks and institutional infrastructure. Unlike those in East Asia, these governments did enact competition laws, but poor policy implementation and inadequate administrative capacity meant that most of these economies remained dominated by entrenched interest groups and monopolistic market structures. The economic landscape in several African and Asian countries tends to be dominated by a few large family or conglomerate interest groups.

Paul Krugman (1994, 1998) has described the East Asian economies as "input" driven, based on high savings rates, good education, and (initially) surplus low-cost labor. Although mobilization of resources was impressive, Krugman viewed it as a one-time effect that was unsustainable without concomitant increases in productivity. And resource allocation was determined primarily by government-directed investments to targeted firms and industries and through special tax and regulatory breaks rather than by market forces. In the long run the costs imposed on current and future generations would be substantial—and unsustainable. Other outcomes included lower levels of private capital formation relative to public investment, diminishing rates of return, and declining total factor productivity (Khemani and Meyerman 1998, Pomerleano 1998, and Young 1993). Krugman (1994) characterized these economies as based on "perspiration" rather than "inspiration" and on "deferred gratification," sacrificing current satisfaction for future gain. Such a model is not necessarily bad in itself, but if inefficient and high-cost structures are insulated from competitive pressures and persist over time, subsequent generations will have to pay the price.

Trade liberalization may not be a substitute for domestic competition law policy (Khemani and Dutz 1996). While East Asian and some other developing economies have generally been considered "open," domestic markets are often highly concentrated with monopolistic market structures and corresponding business conduct and performance. Manufactured goods aimed for export markets also tend to be concentrated in a few sectors with limited upstream and downstream linkages and spillovers across the economy. Preferential credit for selected firms and industries has contributed to concentration and barriers to entry, enabling firms to practice price discrimination between domestic and international markets, with domestic consumers paying higher prices—in part to finance export expansion. Most Korean export firms, for example, have sought to maximize growth or market share rather than profits.

In addition, policy-based barriers to entry and trade and restrictive business practices by firms have constrained both domestic and international competition. Many goods and services are nontradable (for example, high-weight-to-value products such as cement; perishable commodities; and real estate, legal, financial, and various other services), and even for tradable products, international trade does not necessarily provide sufficient competitive discipline in domestic markets. In many economies incumbent firms impede market access through control of raw materials or other key inputs, domestic and international cartels, and exclusive dealing and other restrictive distribution arrangements. Weak domestic competition also distorts the information content of price and profit signals and increases the likelihood of misguided investment-output decisions. Illusionary high profits and low capital costs may encourage firms to overinvest, creating excess capacity.[2]

Consequences of Ineffective Competition: Single, Double, or Multiple Jeopardy?

Most developing and transition market economies, including the once high-performing East Asian economies, have several structural, institutional, and governance characteristics in common:

- *High levels of domestic market concentration, barriers to entry and trade, and low degree of interfirm rivalry.*[3] Even though market liberalization is spreading, these features tend to change slowly because of the weight of past government policies and interventions such as industrial policy, tariff protection, licensing, and preferential procurement, as well as the inherent structure of the economy such as relative size of the domestic markets and underdeveloped capital markets.
- *Lack of an effective market for corporate control—the process of replacing inefficient firm management through mergers and acquisitions.* This is due in part to policy-based restrictions against foreign ownership and in part to high levels of ownership concentration.[4]
- *High levels of ownership concentration and an inadequate corporate governance regime.* In many developing and transition market economies, major corporations are family owned or controlled by a small group of influential investors (Claessens, Djankov, and Lang 1998; Prowse 1998).

These characteristics tend to reinforce one another, creating inflexible and inefficient industrial and financial market structures.

To understand the influence of competition on governance at the state and corporate levels, it helps to look at the risks associated with markets where competition is restricted. High levels of product market concentration and weak competition generally result in high (monopoly) prices and profits. With easy, if not assured, profits and preferential treatment, incumbent firms have little or no incentive to use resources efficiently. Firms insulated from competition generally incur costs higher than those operating under the best technical and managerial practices. Over time, these losses are compounded by the misallocation of resources and by x-inefficiencies stemming from monopolistic output levels and from managerial and organizational slack. Firms may still post satisfactory operating and financial results but only because high prices mask high costs and poor business and investment decisions.

High profits also provide increased incentives for ownership concentration. Why would owners give up ownership control when, on a risk-adjusted basis, they can earn higher returns in markets insulated from competitive pressures?

It is often argued that ownership concentration minimizes the principal-agent problem of ensuring that corporate managers (agents) act in the best interests of the owners (principals) by maximizing shareholder value. This argument has greater validity when ownership and management are separated. If owners are also active managers of the firm, they may exploit minority shareholders, pass the risks and costs on to others (moral hazard), and pursue such noneconomically viable objectives as catering to their own prestige and egos. Ownership concentration also entrenches owner-managers and limits the ability of the market for corporate control to act as a disciplinary force on their performance.

Regulatory barriers and firm-level practices have also tended to limit the scope of competition in takeovers, divestitures, and privatizations in industrial and developing countries alike. The experience of more advanced market economies shows that as regulatory barriers are imposed on corporate control transactions, managerial efforts and board supervision slacken. Firms tend to postpone addressing business problems. Corporate performance generally declines, with adverse consequences for shareholders. During the late 1980s, for example, sharply intensified antitakeover regulations brought contested control transactions to a halt in the United States. A study of the corporate sector during this period found that many leading firms failed to produce any economic value added (the increase in shareholder value less the cost of capital) for their capital and research and development expenditures (Jensen 1993). Although many firms produced satisfactory results despite the deteriorating business environment, the average return on invested capital was surprisingly low.

In developing and transition market economies, impediments to entry and changes in ownership control are often presented as serving the public or national interest. There tend to be restrictions on foreign ownership, reserved areas of economic activity, and licensing requirements and other measures limiting entry. Not only do these policies make it difficult to change ownership and control of a corporation through contested takeovers, they also undermine the development of a cadre of professional independent managers since owner-managers tend to appoint relatives to key management positions.[5]

The commercial advantages of large, closely held incumbent firms are not lost on banks, which tend to dominate financial intermediation in developing countries. Banks maintain cozy relationships with established and well-connected businesses—a natural outcome in a protected and profitable business environment in which both borrowers and lenders operate. In some countries commercial firms also own and control major domestic banks, creating business conglomerates with in-house sources of easy financing for themselves. And bank lending is often determined by political directives, which generally favor large incumbent firms.

Some of these practices contributed to the high leverage in leading firms in East Asia and to the widespread corporate distress and banking failures during the financial turmoil of the late 1990s. More generally, preferred access to bank credit significantly reduces the need of incumbent firms to rely on securities markets, where external financiers demand transparency and accountability of corporate insiders.

Owners of incumbent firms, which have an incentive to retain control of profitable domestic operations, may choose to remain private or to go public without giving up control by retaining a controlling stake or issuing nonvoting shares. Available data suggest that in less competitive markets a higher share of leading firms remains private and that even among publicly traded companies a higher share of firms is closely held (Khemani and Leechor 2000). While concentrated ownership in individual firms is not necessarily a reason for concern, it presents a greater risk of abuse by corporate insiders—especially when firms are insulated from competitive pressures. Unless this risk is mitigated, it is difficult to attract minority shareholders and domestic and foreign investors. Taken to the extreme, ownership concentration and the reliance on internal resources can undermine the development of securities and capital markets.

Economic Power and Political Influence

Regulatory and private restraints on the competitive process have even deeper ramifications. Because established firms tend to be relatively large in size and few in number, they have definite organizational and financing advantages in influencing the government's legislative and regulatory agenda. In more advanced countries where there is a depth of informed opinions, competing interests, and independent media, powerful commercial interests may not always prevail. But in most developing countries competing opinions are more limited. In this context, interest groups are more likely to succeed in furthering their own agendas.

The close connection between economic power and political influence is generally recognized. The successful resistance of public enterprises to privatization programs has been encountered over a wide spectrum of cultural and economic environments, ranging from Ghana to India and Thailand. Another example is the successful opposition of domestic bankers in many countries to competition from foreign banks. Even under the stress of a crisis, major conglomerates in East Asia were able to water down unfavorable reforms and delay implementation.

Incumbent firms often use their political influence to entrench management and corporate insiders. In many jurisdictions incumbent firms can freeze out minority shareholders at unfavorable prices, dilute their voting power by issuing new shares in private placements, or use "poison pills" and other devices that allow them to reject takeover bids without shareholder approval. Despite the obvious risk to investors, change is not easy to effect.

The ability of the corporate elite to resist policy reform is cause for concern. For one thing, inadequate competition limits the access to capital by new or small businesses. Lenders and investors understandably prefer more established firms with significant business advantage. Over time, industrial structure may become skewed, with a few large conglomerates dominating the economy and a large number of small firms struggling with poor prospects for growth. Another concern is that, with distorted prices guiding business decisions, the pursuit of profit may be detrimental to social welfare. Profitable operations based on domestic prices may actually be loss makers when the inputs and outputs are valued at world prices. This has been the case with many commodity monopolies in Africa and politically connected conglomerates in East Asia.

How Competition Can Boost Corporate Governance

In a competitive environment firms cannot expect to earn excess profits. An industry that generates above-average profits tends to attract competitors, and additional supply drives down profitability. Where natural barriers to entry are high, excess profits may persist, and interim regulation may be needed to protect consumers. Over time, however, technological advances and entrepreneurial innovations tend to chip away the natural barriers, unless prevented by regulations.

To withstand competition, firms need operational efficiency. Unless their production and administrative costs are kept below prevailing market prices, which may be determined by efficient competitors at home or abroad, firms cannot service their debt and meet shareholders' expectations. Investors need to evaluate the viability and cash flows of projects, rather than relying on preferential treatment or market power. Under effective competition, preferential treatment can be quickly detected and brought to light by those who suffer the adverse consequences.

Where competition is intense and global, more firms realize that corporate governance makes good business sense. Investors seek out firms that run their business efficiently, treat shareholders equitably, and comply with high standards of disclosure, even when voluntary. Good governance can earn a firm a good reputation and efficient access to finance, enhancing its ability to compete. In effect, good governance becomes an instrument of competitive strategy.

Ultimately, the key role of competition is to enhance economic freedom. It provides opportunities for new entrepreneurs and firms to compete on their economic merits rather than their ability to garner political favor. More business ideas get to face the market test.

Monopolistic market structures, ownership concentration, and lack of an effective market for corporate control are not only reinforcing, but they also lead to other negative consequences, including

- Lobbying, rent seeking, higher barriers against competition from "outsiders" and "foreigners," close government-business relations, and opaque decisionmaking.
- Corruption, bribery, cronyism, and nepotism.[6]
- Problems of governance, at national and corporate levels, and lack of accountability and transparency.
- Underdeveloped capital markets, dependency on bank intermediation, financial-industrial conglomeration, conflicts of interest, and non-arms'-length transactions (Park 1998; Prowse 1998).
- Disenfranchisement of individuals and segments of society from participation in the economy.
- Socioeconomic and political unrest and instability.[7]

In essence, a protected business environment that is intended to provide incentives for capital formation jeopardizes economic development in several ways. These negative consequences stem almost exclusively from the failure to promote effective competition. Policies aimed at fostering a market for corporate control, good corporate governance, greater accountability and transparency, and reduced corruption will fail unless competition in markets for goods and services is promoted. Unlike government directives, the informational content of market-determined price and profit signals provides impartial and efficient benchmarks for gauging corporate performance and redeploying resources from lower- to higher-value uses. Moreover, in a competitive business environment it is not easy to earn excess profits or to pass on high costs to consumers. And extra resources are not available for engaging in corruption, bribery, lobbying, political influence peddling, and rent seeking.

The extent of these influences varies across developing and transition market economies, but anecdotal, case-specific, and empirical data support this general scenario. Economies such as Chile, Hong Kong (China), Singapore, and Taiwan (China) seem to have fared better during the recent economic crisis because of greater domestic and international market competition and more accountable and transparent economic systems.

Designing and Implementing Effective Competition Law Policy

Effective competition law policy interfaces with a broad range of economic policies affecting competition in local and national markets, including privatization, financial markets, international trade, foreign direct investment, intellectual property rights, and sectors in which natural monopolies are likely to occur, such as transportation, power, and telecommunications. Depending on its design and implementation, competition law policy can influence which markets are accessible to firms as well as their pricing, output, and other business strategies.

Opinions differ, however, on whether countries need specific competition legislation to complete their national economic policy framework. Some economists believe that liberalization of international trade is sufficient to promote competition.[8] Others argue that even if competition law is desirable in theory, the cost of such legislation—high probability of improper enforcement, misuse of bureaucratic power, regulatory capture—is so great in developing and transition market economies that it outweighs the possible benefits (Godek 1992). In addition, policymakers in several developing and transition market economies doubt the appropriateness of competition law policy at their stage of economic development, fearing that such policies will disrupt the domestic economy by encouraging the entry of large, powerful multinational firms that will drive out local producers lacking the resources and capabilities to compete with them.

True, competitiveness can be promoted without competition legislation, and having competition laws on the books does not necessarily lead to increased competition and competitiveness. However, it is increasingly recognized that without effective competition law policy, competition and its benefits cannot be safeguarded. Effective competition law policy fosters the development of a market-based economy by preventing unfair and anticompetitive business practices—by domestic or foreign firms—aimed at monopolizing economic activity. Without specific competition law, such practices cannot be addressed. For example, Brazil's competition authority reviewed and modified the terms of several proposed mergers and acquisitions—between domestic manufacturers, between domestic and foreign firms, and between two foreign companies with subsidiary operations—that would have had adverse effects on competition. Moreover, in several developing and industrial countries there have been cases involving cartels of domestic and international firms that have engaged in price fixing of inputs, raising the domestic costs of production and undermining the competitiveness of firms.

Competition law policy can be introduced gradually, with an adjustment period to allow firms to adapt to the new business environment. A gradual approach has been the practice in several industrial countries, such as Canada and the United Kingdom, which have made major changes or enacted new competition legislation. The period of adjustment should be limited, with an irreversible termination date. If firms cannot adjust to the new business environment in a reasonable period of time—say, five to seven years—their management competence must be questioned. Why should domestic consumers have to pay higher prices so that inefficient producers can continue to reap excess profits?

The views of economists, lawyers, and practitioners have converged on the principal elements of best practice for maintaining and encouraging competition. These elements are particularly relevant to developing and transition market economies, with their histories of trade protection, central planning, a large role for state enterprises and other government interventions, and high levels of industry and aggregate concentration. These principles of best practice also lower the risks of misuse of bureaucratic power and capture, a frequent concern of critics of competition law policy.

Policy Approach for Business Conduct and Business Environment

The principal objective of competition law policy should be to maintain and encourage competition for promoting economic efficiency and maximizing consumer welfare. It should aim at protecting the process by which firms compete, not competitors. What matters is the actual or potential business conduct of firms in a given market—whether a firm (or group of firms) can engage in business practices that substantially lessen competition—not industry concentration or the absolute or relative size of firms. Firms can exercise market power only when existing or potential competition is inadequate to constrain such behavior, indicating that barriers impede new entry and competition.

This fact means that competition law policy needs to focus not only on the business conduct of firms, but also on the business environment in which firms operate. And that is influenced by other government economic policies. As the experience in several developing and transition market countries suggests, regulatory, trade, investment ownership control, and industrial policies have adversely affected competition. Thus these policies are potential instruments for maintaining and encouraging competition along with the substantive provisions of competition law. The two sets of policies need to be harmonized and linked. Altering the business environment in this way to promote competition inculcates sound business practices and ethics as well as constraining anticompetitive behavior. By enhancing the mobility of resources, competition facilitates structural adjustment and reduces the need for costly direct government economic interventions and extensive enforcement of competition law.

Scope of Competition Law

Competition law should be a general law of general application to all sectors of the economy and to both public and private sector firms engaged in commercial economic activity. No exemptions should be granted except those aimed at facilitating legitimate economic activity. Traditional examples of legitimate exemptions are collective bargaining by labor to negotiate terms and conditions of employment, collective underwriting of risk by insurance companies, the primary issuance of new stock by investment brokers, and dissemination of information and development of product standards by trade associations. Permitting firms to underwrite risky businesses or new stock issues collectively does not mean that they should not compete vigorously against each other on the prices and services they provide. Including state enterprises under competition law is particularly important in the transition market economies, where privatization remains incomplete.

Competition law should include provisions explicitly prohibiting business practices that reduce economic efficiency and consumer welfare. In some jurisdictions some practices (such as agreements between firms to fix prices, rig bids, restrict output and market shares, and allocate geographic markets or customers) are classified as illegal and are subject to severe penalties. Other types of business practices may not have such clear anticompetitive effects, and some may even increase efficiency and consumer welfare. These need to be assessed under a *rule-of-reason* approach that weighs their relative costs and benefits. Examples would be horizontal and vertical mergers, specialization agreements, joint ventures, and various types of vertical manufacturing and wholesale-retail distribution arrangements.

These provisions have to be carefully drafted and administered to ensure that the law does not unduly constrain legitimate business conduct and introduce uncertainty into the market. In addition, clearly permissible practices need to be identified, such as interfirm cooperation in the development of standards, collection and exchange of industrywide statistics, research and development, and certain types of joint ventures. To enhance business confidence and clarify policy objectives, appropriate enforcement guidelines need to be developed and publicized.

Competition law generally contains two types of provisions, conduct and structural.

CONDUCT PROVISIONS. The conduct provisions of competition law relate primarily to

- Horizontal agreements between firms to fix prices, engage in bid rigging, restrict output or market shares, and allocate geographic markets or customers.
- Abuse of market position by large dominant firms, such as predatory price and nonprice strategies.
- Vertical restraints between suppliers and distributors such as resale price maintenance, exclusive dealing, and geographic market restriction.

Economic theory suggests no efficiency or consumer welfare reasons for horizontal agreements between firms to limit price or output competition. Firms engaging in such behavior interfere with the competitive process and reduce the benefits that flow to society from competition. Economists and lawyers in all jurisdictions agree almost unanimously that such agreements should be strictly prohibited and subject to heavy penalties. Canada, the United States, and many other industrial countries view price-fixing agreements and other types of collusive arrangements as criminal behavior and prosecute violations under the penal code. In many developing and transition market economies, a history of price controls and industry concentration tends to foster such behavior, making it important that prohibition of the behavior be clearly articulated.

While large relative size in a market should not itself be illegal, the abuse of a dominant market position needs to be prevented because it substantially lessens competition. Distinctions should be drawn between maintaining dominance through superior competitive performance (more efficient production, higher quality products or services, better customer service) and entrenching market position through such anticompetitive practices as preemptive purchase or control of vital input sources or distribution facilities and predatory pricing to drive out or discipline competitors. Since the distinction is not always clear-cut, a rule-of-reason approach that examines firm behavior in terms of probable business intent and anticompetitive impact is often advocated for the administration of abuse of dominance provisions.

Economists and lawyers are divided on firm motivations for vertical restraints and on their economic implications. However, an increasing number of practitioners advocate a rule-of-reason approach on the grounds that vertical restraints are more likely to affect competition adversely if the firms involved have dominant market position and if large barriers to entry exist. This view suggests that no specific provisions on vertical restraints are necessary. Where facts warrant, the cases can be examined under the provisions of competition law that govern abuse of dominance.

STRUCTURAL PROVISIONS. The primary structural provisions of competition law relate to mergers, acquisitions, and joint ventures. The principal competition concern is with horizontal mergers and acquisitions. One view is that when such transactions significantly reduce the number of independent firms and increase concentration in the market, substantial lessening or prevention of competition may occur. Another view is that these transactions are generally motivated by efficiency concerns and that substantial lessening of competition can occur only if there are barriers to entry. Competition policy should therefore focus on reducing or eliminating these barriers. Both sides generally advocate a rule-of-reason approach for evaluating mergers.

Since horizontal mergers and acquisitions may both substantially lessen competition and increase economic efficiency, a cost-benefit approach in implementing merger policy is advisable. Some jurisdictions exempt mergers or permit them to proceed on a restructured basis if the gains in economic efficiency (and other benefits in the public interest) are likely to be greater than the losses from reduced competition. This policy facilitates structural adjustment and more efficient use of resources because it allows for the closure of suboptimal plants through mergers and acquisitions. Canada, the European Union, Germany, Japan, and several other jurisdictions take a similar approach to joint ventures, permitting firms to rationalize production lines and enter into specialization agreements to achieve longer production runs and realize potential efficiencies.

Mergers, acquisitions, joint ventures, and specialization and rationalization agreements can irreversibly alter the structure of industry. Prior notification and approval of such business arrangements is thus recommended. To avoid imposing an unnecessary regulatory burden on firms, only the largest transactions or proposals should be screened. A common approach specifies size thresholds in terms of market share, assets, sales, or employment of the parties involved. Firms are required to file relevant business information, and strict time periods are set for expeditious review and clearance of cases. The extensive restructuring and privatization of state enterprises in transition market economies suggest that appropriate implementation of these structural provisions of competition law is especially important.

Administration and Enforcement of Competition Law

Jurisdictions such as Canada and the United States employ a combination of criminal, civil, and administrative legal procedures in administering and enforcing competition law. Other jurisdictions, primarily in Europe and Latin America, rely solely on administrative legal procedures. Horizontal price fixing or collusive bidding agreements are strictly prohibited under both approaches, while treatment varies for other types of horizontal restraints, such as geographic market allocation, parallel pricing, and restrictions on output. Structural and other provisions of competition law to which the rule of reason is applied are better suited to civil and administrative procedures. Since civil courts and administrative tribunals in most transition market economies are unlikely to have experience in competition-related matters, they might consider setting up specialized administrative bodies or tribunals to enforce competition law, drawing on experts in law, economics, finance, and various fields of industry and business.

Whatever institutional mode is adopted, certain principles are important to the design of an effective office of competition law:

- The competition policy agency should be independent and insulated from political influence. It should be accountable to an impartial body, such as parliament.
- The investigation-prosecution function should be separated from the adjudication function.
- A system of checks and balances should be in place, with rights of appeal of decisions on legal and economic grounds.
- Cases should be resolved expeditiously to avoid unnecessary transaction and other business-related costs.
- Proceedings should be transparent while safeguarding sensitive business information of a competitive nature.
- Measures for collecting market- and firm-specific information need to be in place.
- The proceedings should be accessible to all affected parties, with provisions for introducing expert testimony and evidence.
- Provisions should be made for imposing fines and other remedial measures to deter or correct infractions of the law.

The competition law policy agency should also have a statutory role in formulating and commenting on government economic and regulatory policies affecting competition in the marketplace (Box 4.1). Such a competition advocacy role could counter or minimize the adverse effects of rent-seeking behavior prevalent in most countries. Because of the limited administrative capacity and enforcement experience in this field in most developing and transition economies, some analysts view this role as the most important—if not the sole—function of a competition policy agency (Kovacic 1995, 1997; Rodriguez and Williams 1994). Competition advocacy can also help prevent the misapplication of provisions of competition law, which could introduce further distortions into the economy.

The burden of implementing competition law does not have to rest solely with the government. Allowing individual and class action law suits strengthens application of the law and promotes a competition-oriented culture. Safeguards are needed to prevent the abuse of these provisions and to avoid excessive litigation. If private actions are allowed, the law does not have to be administered by a large bureaucratic office. The specialized court can consist of few full-time members and draw on a roster of part-time officials as a need for specific expertise arises. Implementation of the law can also proceed in stages, starting with competition advocacy and the prohibition of price fixing and other collusive agreements between firms. In the more complex areas of competition law, such as abuse of dominant market position and mergers, only the most blatant infractions need be pursued until the necessary experience for more subtle cases has developed.

Constraints on Effective Implementation of Competition Law Policy

The greatest impediments to effective implementation of competition law policy are a lack of political commitment and the political and economic influence of entrenched interest groups. Implementation of competition law policy also requires highly specialized knowledge and expertise—often lacking in developing and transition market economies because of their economic legacy and current economic circumstances:

- Government interventions and central planning have resulted in industrial structures with high degrees of concentration, state ownership, and controls; distorted economic incentives; segmented markets; a scarcity of entrepreneurship and managerial skills; and weak or inadequate financial capital markets.
- Economic liberalization has revealed severe rigidities and bottlenecks in the movement of resources from lower- to higher-valued uses. Inflation rates, while lower than they once were, are still high in some transition

Box 4.1 Competition Advocacy in Promoting Competition in Government Policies

Not only do private restrictive business practices lessen competition, so too may public policies and institutional arrangements. Indeed, private restrictive business practices are often facilitated by government interventions in the marketplace. Through competition advocacy—advising, influencing, and participating in government economic and regulatory decisionmaking to promote more competitive industry structure, firm behavior, and market performance—competition offices can become a voice for policies that interfere least with market forces. Three areas are particularly important:

- Industrial policy, including trade policy.
- Economic regulation.
- State enterprise operations and privatization.

In Canada, for example, high levels of tariff protection (before the North American Free Trade Agreement) facilitated price-fixing agreements in the plate glass, fertilizer, pharmaceuticals, and sugar industries.

Economic regulation of sectors thought to be natural monopolies, such as electricity and telecommunications, has been implemented through vertical integration and exclusive licenses for products and services for which effective competition is possible, such as equipment supply, electricity generation, and long-distance telephone service. In Australia and New Zealand competition authorities oversee the business practices of infrastructure companies; in other jurisdictions sector-specific regulatory authorities do this work. Although no one approach can claim superiority, the risk of regulatory capture is greater with sector-specific agencies, making it important that competition law policy principles be embedded in sector-specific regulations. Provision should be made for the competition authority to participate in regulatory economic decisionmaking, to ensure a voice for pro-competition considerations.

Governments are often caught in a conflict of interest when it comes to state enterprise reform and privatization. In a desire to maximize revenue from the sale of state assets, governments may end up transferring public monopolies to private ones. This was alleged to be a factor behind the sale of British Caledonia airlines to British Air instead of to SAS; the sale prevented the introduction of new competition into the U.K. holiday charter market. The acquisition of Skoda by Volkswagen in the Czech Republic was reportedly accompanied by demands for "incentives" in the form of tariff protection and restrictions on foreign investment, which would limit import competition and new entry. In Jamaica the telecommunications company was privatized with exclusive rights for a period of 25 years.

Competition advocacy can help to prevent such behavior—or at least subject it to greater accountability, transparency, and public discussion.

market economies. As a consequence, price signals tend to mask or distort the real profitability of investment opportunities.

- Weak legal and institutional frameworks and inadequate skills base and administrative capacity impede the work of competition authorities. In many transition market economies, competition offices are staffed with people trained in physical sciences and engineering instead of economics and finance, or who are experienced in administering price controls instead of competition policy.

- A lack of political will and inadequate budgets undermine support for effective implementation of competition law policy. Because competition does not discriminate in that it does not favor one constituency over another, there are few political or business constituencies with a vested interest in promoting and building a competition culture.

These and other impediments can be addressed through technical assistance and other forms of aid. At the same time, the increasing international competition faced by firms and national economies in attracting investment and integrating into the global trading system to gain market access are providing further incentives to implement competition law policy.

Concluding Remarks

Promoting effective competition is critical for fostering sustainable economic development. By maintaining and encouraging competition, governments in developing and transition market economies can address several of the factors that discourage investment and competitiveness. These factors include the risks associated with high levels of industry concentration and monopolistic market structures, such as high prices, high costs of production, and low levels of operational efficiency; concentrated corporate ownership structures and poor corporate governance, such as abuse of shareholder rights, conflicts of interests, insider dealings, and misguided investment decisions; the lack of accountability and transparency in government-business relations; and opportunities for corruption and bribery, lobbying, and rent-seeking behavior.

To counter such tendencies, countries need credible competition law policy, which seeks to foster economic efficiency and consumer welfare by addressing both unnecessary public policy and private restraints on competition. The design and implementation of this policy should contain checks and balances and measures to ensure accountability and transparency in administration.

Effective implementation of competition law policy is not easy. It requires political will and commitment as well as specialized legal and economic skills and institutional capacity. There tends to be little popular support for competition law policy and considerable opposition from large, politically connected firms and interest groups. However, in an increasingly integrated global economic system, firms have to compete not only in marketing their products, but also in attracting foreign investment. Both governments and large domestic firms have less and less freedom to pursue policies that limit competition and favor self-serving interests. Doing so will delay the efficient restructuring of their economies and undermine their chances for improved industrial competitiveness.

As a starting point, competition law policy should foster interfirm rivalry and competition in domestic markets. This step will increase competitiveness, easing adjustment and integration into the global economy. If firms do not compete in their home markets, they are less likely to be competitive in international markets.

Notes

1. Various definitions of corporate governance have been put forward. As used here, the term refers not only to the relationships among corporate managers, directors, and providers of equity capital, but also to the relationship of the corporation to other stakeholders and society. It includes the laws, regulations, stock exchange listing rules, and voluntary private sector practices that enable the corporation to attract capital, perform efficiently, and be held accountable so as to maximize its value (Gregory 2000).

2. Firms in the Republic of Korea, for example, built up costly excess capacity, primarily by using extended credit lines instead of internally generated funds. The close relations between industrial companies and financial institutions, including government-owned banks, led to reliance on bank credit instead of the capital market for financing growth. Korean companies had (and many continue to have) extraordinarily high debt-equity ratios—in many cases exceeding 400 percent. This situation is not unique to Korea; other countries, including Indonesia, Mexico, and Thailand have had similar problems.

3. For example, the top 30 Korean *chaebol* (conglomerates) dominate every sector of the economy except agriculture. They own or control roughly two-thirds of the 100 largest manufacturing firms and before the crisis accounted for about 40 percent of manufacturing GDP (gross domestic product), 16 percent of total GDP, 50 percent of exports, and 15 percent of commercial bank loans (Min 1998; Min and Jae 1997). In Indonesia former President Suharto and his extended family, along with favored politicians and business people, controlled virtually every significant financial and industrial sector in the economy. High levels of industry and aggregate concentration also exist in Russia (Ryterman 1999) and in other developing countries, such as Malaysia, Pakistan, the Philippines, and Turkey.

4. Several East Asian crisis countries restricted foreign ownership and contested (hostile) takeovers. In many spheres of economic activity, firms were required to obtain government permission or licenses to enter into or conduct business.

5. In Korea 29 of the 30 *chaebol* were (and most still are) family controlled. In many cases *chaebol* owners are interlinked through marriage, interlocking directorates, cross-holdings, and investments as well. Marriages also cemented links with senior government officials. Cummings (1997) points out that the sons or daughters of one-third of *chaebol* owners were married to high-ranking government officials.

6. In Korea liberalization and democratization have led to prosecution and imprisonment of several senior government officials on charges of corruption—including former presidents. Actions have also been recently initiated against President Suharto and some members of his family in Indonesia and against officials in India, Thailand, the Philippines, and other countries.

7. Unlike Indonesia, Korea has been able to control some of these unfortunate outcomes of economic-financial crises. However, there have been notable demonstrations and protests against government policies and measures and against the *chaebol*.

8. Counterarguments can be found in Khemani and Dutz (1996). Moreover, in recent years international cartels have operated within and across the borders of Canada, Western Europe, and the United States, all among the most open markets in the world. The products covered include various inputs such as aluminum phosphide, bromine, citric acid, graphite rods, and lysine, as well as more consumer-oriented products such as sugar, plastic dinnerware, and vitamins. The competition authorities have fined companies involved in illegal price-fixing conspiracies in amounts reaching more than $1 billion (Evenett, Levenstein, and Suslow 2001).

References

Claessens, Stijn, Simeon Djankov, and Larry H. P. Lang. 1998. "Who Controls East Asian Corporations?" Policy Research Working Paper 2054. World Bank, Economic Policy Unit, Finance, Private Sector, and Infrastructure Network, Washington, D.C.

Cummings, B. 1997. *Korea's Place in the Sun: Modern History.* New York: W.W. Norton & Co.

Evenett, S., M. Levenstein, and V. Suslow. 2001. "International Cartel Enforcement: Lessons from the 1990s." World Bank, Development Research Group, Washington D.C. Processed.

Godek, P. E. 1992. "One U.S. Export Eastern Europe Doesn't Need." *Regulation* 15 (1).

Gregory, H. 2000. "The Globalization of Corporate Governance" *Global Counsel* 5: 52.

Jensen, M. C. 1993. "The Modern Industrial Revolution: Exit and the Failure of Internal Control Systems." *Journal of Finance* 48: 831–85.

Khanna, T., and K. Palepu. 1999. *Emerging Market Business Groups, Foreign Investors, And Corporate Governance*. NBER Working Paper 6955. Cambridge, Mass. National Bureau of Economic Research.

Khemani, R. S. 1994. "Competition Law." Viewpoint 14. World Bank, Private Sector and Infrastructure Network, Washington, D.C.

Khemani, R. S., and M. Dutz. 1996. "The Instruments of Competition Policy and Their Relevance for Economic Development." Private Sector Development Occasional Paper 26. World Bank, Private Sector Development Department, Washington, D.C.

Khemani, R. S., and C. Leechor. 2000. "Competition Boosts Corporate Governance." *Global Competition Review* (February/March).

Khemani, R. S., and G. Meyerman. 1998. "East Asia's Economic Crisis and Competition Policy." *Global Competition Review* (August/September): 16–18.

Kovacic, W. E. 1995. "Designing and Implementing Competition and Consumer Protection Reforms in Transitional Economies: Perspectives from Mongolia, Nepal, Ukraine, and Zimbabwe." *De Paul Law Review* 44: 1197–1224.

———. 1997. "Getting Started: Creating New Competition Policy Institutions in Transition Economies." *Brooklyn Journal of International Law* 23: 403–53.

Krugman, Paul. 1994. "The Myth of Asia's Miracle" *Foreign Affairs* November/December).

———. 1998. "Asia: What Went Wrong?" *Fortune* 137 (4): 32.

Min, Yoo Seong. 1998. "The *Chaebol* under Fire: The Real Issues in Corporate Restructure. Korea and the Asian Economic Crisis: One Year Later." Paper presented at a symposium sponsored by Georgetown University, Korea Institute of America, and the Korea Institute for International and Economic Policy. Processed.

Min, Yoo Seong, and Lim Young Jae. 1997. *Big Business in Korea: New Learnings and Policy Issues*. Seoul: Korean Development Institute.

OECD (Organisation for Economic Co-operation and Development). 1999. "Structural Aspects of the East Asian Crisis." OECD Centre for Co-operation with Non-Member Countries, Paris.

————. 2000. "Review of Regulatory Reform in Korea." Paris.

Pomerleano, Michael. 1998. "The East Asia Crisis and Corporate Finances: The Untold Micro-Story." Policy Research Working Paper. World Bank, Development Prospects Group, Washington, D.C.

Prowse, S. D. 1998. "Corporate Governance in East Asia: A Framework for Analysis." Paper prepared for conference on "Managing Capital Flows: National and International Dimensions," June 15–16, Bangkok, Thailand. Processed.

Rodriguez, A. E., and M. D. Williams. 1994. "The Effectiveness of Proposed Antitrust Programs for Developing Countries." *North Carolina Journal of International Law and Commercial Regulation* 19: 209.

Ryterman, Randi. 1999. "Russian Law Reform." *McGill Law Journal* 44 (2):255-537.

Stone, A., K. Hurley, and R. S. Khemani. 1998. "The Business Environment and Corporate Governance: Strengthening Incentives for Private Sector Development." World Bank, Private Sector Development Department, Washington, D.C. Processed.

Young, A. 1993. "Lessons from the East Asian NICs: A Contrarian View." Working Paper 4482. Cambridge, Mass.: National Bureau of Economic Research.

5
Supporting Technology Generation and Diffusion at the Firm Level

Manjula Luthria

Strengthening the technological capability of firms is critical to making the transition from low-wage, low-skill jobs to higher value-added activities capable of contributing to economic growth and poverty reduction. A necessary ingredient of industrialization, developing strong technological generation, diffusion, and absorption capabilities is the key to improving firm-level productivity. This chapter focuses on incentives and institutions that impede or help firms upgrade their technologies.

Why Countries Cannot Afford to Miss the Boat

A number of factors are at play that help set the context in which firms in developing countries need to approach their technology-upgrading efforts.

Patterns of World Trade

Two important trends in world trade make it urgent that countries build and leverage their technology capabilities. Technology-based products are gaining importance in world trade, and developing countries are increasingly contributing to this trend.

Growth in manufactured exports is fastest for high-technology products not only for the world as a whole, but for both industrial and developing countries alike. Developing country shares of technology-based products have increased substantially, especially in electronics and other high-technology products (Table 5.1).

Disparities in Research and Development Effort

Industrial countries account for nearly all research and development (R&D) activities (Table 5.2). If R&D is taken as a proxy for knowledge creation

Table 5.1 Manufactured Exports by Technological Subcategories, 1985–95
(percent)

Manufactured export	World	Annual growth rates 1985–98 Industrialized countries	Annual growth rates 1985–98 Developing countries	Developing world shares 1985	Developing world shares 1998
All manufactures	9.7	8.8	12.5	16.4	23.3
Resource-based	7.0	7.0	6.0	26.3	23.7
Low-technology	9.7	8.5	11.7	26.7	34.5
Medium-technology	9.3	8.5	14.3	8.3	15.3
High-technology	13.1	11.3	21.4	10.7	27.0
Electronic	14.1	11.7	22.1	14.0	34.2
Other	11.0	10.7	16.1	4.8	8.6

Source: Lall, 2000.

Table 5.2 Research and Development Effort across Countries

Country	As a percentage of GNP	Spending per capita ($)	Country	As a percentage of GNP	Spending per capita ($)
Canada	1.6	310.1	India	1.1	3.7
France	2.3	574.8	Pakistan	0.3	0.8
Germany	2.3	632.7	Sri Lanka	0.2	1.4
Japan	3.0	1,189.2			
United States	2.5	674.5	Brazil	0.4	14.6
			Mexico	0.4	13.3
China	0.6	3.7			
Indonesia	0.2	2.0	Egypt	1.0	6.0
Korea, Rep. of	2.7	261.9	Jordan	0.3	4.5
Malaysia	0.4	15.6	Turkey	0.4	11.1
Philippines	0.1	1.1			
Singapore	1.1	294.0	Mauritius	0.4	13.5
Thailand	0.2	5.5	Nigeria	0.1	0.3

Source: UNESCO, 1995.

efforts, and if knowledge is becoming the chief engine of growth (some have likened it to electricity during the Industrial Revolution), this disparity in knowledge creation may have serious implications for the ability of poor countries to catch up with the richer ones.

Mounting Pressures on Firms

With increased globalization has come a rapid division of the production process, along with outsourcing and subcontracting of important production activities. It is not uncommon for a firm to compete with firms in distant countries for resources and opportunities within its own domestic markets. Firms consequently are pushed to reexamine their activities and refocus on their core competencies, carefully selecting the appropriate technology for each part of the production process. Even firms that do not produce technologies need to master and adapt existing technologies if they are to maintain a competitive edge. Such adaptation is constrained if technological awareness and effort are lacking at the domestic level.

The lesson that emerges is that comparative advantage based on resource endowments or cheap labor is no longer enough to compete. Unless developing countries concentrate on raising technological effort immediately, they risk being caught in a low-wage, low-skills trap.

What Firms Need to Improve Their Technological Capabilities—and Why They Can't "Just Do It"

Firms decide whether to undertake technological upgrading of their activities, but that decision and the effort put into the upgrading are heavily influenced by surrounding conditions. Especially important is an environment that fosters competition among firms, forcing them to reexamine the strengths and weaknesses of their production capabilities. Open trade regimes and appropriate domestic competition policies for entry and exit of businesses provide incentives for firms to develop, improve, adapt, and adopt appropriate technologies.

Also important is the right technology support infrastructure, including research and technology institutions, standards and quality certification laboratories, and testing facilities. Small and medium-size firms, in particular, need to be able to use these institutions in a problem-solving capacity. Universities and vocational training institutes, which supply a pool of skilled labor, are also an important part of this infrastructure.

Even the less developed countries usually possess a few such institutions, albeit with limited scope and coverage. Yet firms face extraordinary difficulties in getting their technology upgrading efforts off the ground. Typically, the biggest constraint is a lack of information about such support institutions, often because these institutions lack a client focus and fail to respond to client demand. Links between firms, universities, and training institutes are often weak, so firms lack the skills needed to develop and deploy appropriate technology. Also missing are mature financial institutions, instruments of long-term finance to firms, and venture capitalists and other institutions that know how to evaluate and fund risky innovation.

Public policy needs to address these informational, institutional, and financial market deficiencies and to work with businesses to provide the three key ingredients of a successful technology development program:

- Building high-quality human capital and skills.
- Strengthening technology infrastructure and its links with industry.
- Improving access to finance for technology development.

Although neither independent nor exhaustive, these activities nonetheless outline an action program for improving the technological capability of firms.

Building High-Quality Human Capital and Skills

A high-quality work force enables firms to respond flexibly to rapid economic and technological change, to produce higher-quality products, to adopt and improve on new production processes and technologies, and to develop new skills as the structure of jobs evolves. Over the past decade, concerns about the supply of skilled workers have become acute in both developing and industrial countries.

For policymakers, the challenge is to develop worker capability through the effective use of public and private sector educational resources and through training incentives for firms and workers. As a start, universities and technical and vocational institutes should align their programs more closely with demand and supply in the market for skilled labor. Such coordination with private industry can take place by sharing costs for certain strategic educational programs and by formalizing internships and other personnel exchange programs.

Firms need to undertake training activities on their own as well. The productivity benefits of training can be quite high. Tan and Batra (1997) and Batra (2001), modeling training in a production function framework using labor, capital, and training as inputs, found productivity increases attributable to training alone of 71.1 percent in Indonesia (1992), 44.4 percent in Mexico (1992), 30.3 percent in Guatemala (1999), 28.2 percent in Malaysia (1994), and 26.6 percent in Colombia (1992).

Yet few firms undertake formal training, for a variety of reasons including being unaware of its importance, assuming that mature technologies can be absorbed costlessly, substituting on-the-job training, and fearing the loss of trained employees to competitors. In an effort to encourage firms to train, Malaysia, Pakistan, and the Philippines, among others, have used tax incentives for subsidizing employee training. Chile, Mexico, and others provide subsidized training through small and medium-size enterprise support centers. In a more market-friendly variation, Brazil, Singapore, and Tunisia use training vouchers, which allow firms to purchase subsidized training from a list of approved providers. In Malaysia and Morocco the vouchers can be used at industry-managed training centers (see Chapter 6 for an evaluation of these and other policies).

Strengthening Technology Infrastructure and Links with Industry

Technology infrastructure varies by country size, capacity, purpose, and sectoral or industry concentration, but generally it can be grouped into three broad categories:

- Measurement, standards, testing, and quality institutions.
- Research and technology laboratories and extension centers.
- Linkage-creation institutions: technoparks, incubators.

Measurement, Standards, Testing, and Quality Institutions

A measurement, standards, testing, and quality system typically includes a keeper of primary measurement standards (length, weight, sound) and a second tier of calibration and testing laboratories. (For example, the scales at airports might be calibrated by a secondary laboratory dedicated to airline services, while the laboratory's equipment would be calibrated by a primary measurements center.) A standards and quality institute would be involved with certification, such as ISO 9000 certification for the production processes of businesses. An accreditation body that certifies these facilities would also be needed. And finally, the system would include facilities for testing the physical and chemical attributes of products and processes.

As firms move from low-skill goods to technology- and skill-intensive products that demand precision and quality, the need for standards and testing infrastructure grows. To engage in international trade, firms need to meet technical standards in many areas. That means that in countries without an adequate standards and testing infrastructure, firms have to send their equipment abroad for calibration and testing and may need to purchase additional machinery to cover this down time. Firms may also incur high transport costs, especially for fragile machinery, and they miss out on the access to local specialists, equipment, and facilities that can help firms improve their products and processes. Small and medium-size firms are particularly disadvantaged because they can neither afford the costs of shipping their equipment abroad for calibration nor engage the services of expensive consultants, thus impeding their ability to comply with international directives (Box 5.1). Investments in Turkey's measurement institute, for example, have allowed domestic firms to access these services at a price nearly three times lower than in neighboring countries (World Bank 1999).

Because many different industries typically use the measurement standards, and because the costs associated with purchasing the calibration equipment, hiring the skilled personnel, and conducting related R&D are beyond the capacity of most firms, standards and testing investments receive public sector support in most countries. When cost reductions to local firms and the spurt to related second-tier services are accounted for, the social rate of return on these investments is high (Box 5.2). In addi-

---

Box 5.1 Technical Barriers to Trade

The Organisation for Economic Co-operation and Development (OECD 1999) conducted a survey of 55 firms in Germany, Japan, the United Kingdom, and the United States to determine how much technical standards and conformity assessment procedures impede trade. The study estimated the additional costs of compliance at between 0 and 10 percent. Some firms, generally the smaller ones, chose not to enter a particular market because of expected regulatory costs, suggesting that standards are relatively greater deterrents to entry for small firms.

The U.S. International Trade Commission (USITC 1998) conducted informal interviews on the same topic with corporate executives and officers of trade associations in the information technology industry in Asia, the European Union (EU), Latin America, and the United States. Many information technology firms viewed testing and certification requirements as substantial barriers to trade. For example, meeting EU tests for telecommunications equipment was estimated to take six to eight weeks and to reduce product value by 5–10 percent. Registering products to recognized standards, such as ISO 9000, cost more than $245,000 per U.S. telecommunications firm.

Henson and others (2000) studied the problems faced by firms in developing countries in meeting the standards imposed by the agreement on the application of Sanitary and Phytosanitary Standards. Such requirements were ranked as the most significant constraint on exporting agricultural and food products to the European Union—ahead of transport costs, tariffs, and quotas.

Source: Maskus, Wilson, and Otsuki, 2000.

Box 5.2 National Institute of Standards and Technology Initiatives

The National Institute of Standards and Technology (NIST), in the U.S. Department of Commerce, is responsible for setting and maintaining physical and chemical standards. Two examples of the economic benefits of their programs follow.

Thermocouple Calibration Program. The Thermocouple Calibration Program (TCP) maintains, improves, and disseminates national standards of temperature by supporting research on the basic properties that underlie the measurement science needed to provide primary calibration services. Benefits to industry were estimated based on surveys and interviews that asked participants to estimate the additional expenses that would be incurred were NIST to cease its services.

Thermocouple Calibration Program Costs and Benefits
(thousands of US$)

Year	NIST costs	Net industry benefits	Year	NIST costs	Net industry benefits
1990	220.4	—	1996	174.7	—
1991	325.9	—	1997	185.2	2,296.8
1992	483.2	—	1998	196.3	2,411.6
1993	266.6	—	1999	208.1	2,532.2
1994	206.1	—	2000	220.6	2,658.8
1995	211.8	—	2001	233.8	2,791.8

— Not estimated.

Measuring sulfur in fossil fuel. The sulfur content of fossil fuels is one of the most important intrinsic factors affecting fuel prices. Environmental regulations have placed increasingly lower limits on sulfur content and impose large fines for noncompliance. These regulations require a determination of the sulfur concentration in fossil fuels by two or more laboratories.

NIST developed a method to determine sulfur content in fossil fuels to an accuracy of better than ± 0.1 percent relative. NIST certified the sulfur content in about 30 coal and fuel oil standard reference materials. These standard reference materials provide the primary calibration materials needed for instrumentation used in routine measurements. Standard reference materials also provide industry with a strong traceability link to NIST for such measurements, whether for setting the price of fuel or for demonstrating compliance with environmental regulations. Surveyed industry representatives indicated that these standard reference materials have decreased the level of uncertainty associated with measurements of sulfur content and have created economic benefits throughout the supply chain. Economic benefits include improvements in product quality and production efficiency, lower transaction costs, and reductions in sulfur emissions. The ratio of benefit to cost was estimated at 113, the social rate of return at 1,056 percent, and the net present value at more than $400 billion.

Source: Semerjian and Watters, 1999.

tion, if measurement institutions were expected to recover all associated costs, many potential users of standards would be discouraged, and companies would be at a disadvantage relative to their international competitors. Cost recovery in primary services ranges from 20 percent to 30 percent, although secondary labs for certification and testing can be profitmaking. Public support is therefore warranted for subsidizing primary measurement institutes and fostering a secondary tier of private labs for delivery of services.

Not all countries need full-fledged measurement institutions. Studies should identify the sectors most likely to need related services, followed by an estimate of which services can be provided in-house and which require sending equipment to neighboring countries, and what the cost-benefit ratio is for domestic investments. Accreditation bodies and standards institutes can share some costs with industry, while remaining government-mandated and -funded in order to ensure impartiality.

Research and Technology Laboratories

Research and technology laboratories funded by central or state governments or affiliated with universities and technology institutes exist in almost every industry—from defense, aerospace, and telecommunications to consumer-related industries such as textiles, food, and beverages. While many of these research facilities are part of large companies that can afford specialized R&D, much of this research and technology infrastructure is funded by the public sector or was started with public funding.

Facilities with public sector ties often lack an industry focus, so domestic industry fails to benefit from the productivity enhancements these laboratories should enable. These facilities are increasingly an unsupportable burden on the public budget and have had to find ways to earn income by selling their services.

Restructuring these research and technology facilities to recover some of their costs is an important reform in countries where such infrastructure is languishing. Reform includes greater responsiveness to the needs of the private sector and better publicizing of available products and services. When India restructured its Center for Scientific and Industrial Research network, income from contractual research work went from zero to as high as 70 percent in some cases as the institutions worked with firms in joint problem solving. And as a consequence of weaning themselves from government support, the institutions have been able to withstand the massive budget cuts that research institutes worldwide have been experiencing in recent years. Making this transition is not easy, and it needs to be done slowly, taking into account the different priorities and approaches of businesses and the scientific community (Box 5.3).

Box 5.3 Barriers to Reform of Research and Technology Laboratories

Research and technology laboratories in the Center for Scientific and Industrial Research system in India underwent massive restructuring in the early 1990s in an effort to make these public institutions more responsive to the needs of the private sector. A questionnaire circulated to 35 laboratories revealed the top five barriers they faced in undertaking collaborative research with the private sector:

- Lack of industry awareness of its own technology needs or of the capabilities of research and technology laboratories.
- Lack of industry confidence in the capabilities or facilities of the laboratories.
- Lack of incentives for scientists to transfer technology to the marketplace.
- Scientists' lack of experience interacting with industry.
- Excessive bureaucracy in public institutions, discouraging industry from seeking partnerships to work jointly on projects.

Worries about confidentiality and a sense that labs and industry were competitors were rated as the least important barriers to collaboration in this survey, while anecdotal evidence gathered from firms reveals these concerns to be formidable obstacles to collaboration.

When restructuring, research and technology laboratories should thus pay special attention to:

- *Allocating resources to improve interaction.* The key seems to be addressing some of the information market failures that impede collaboration. One way to do this is to designate a facilitator of industry-firm interactions who would be responsible for making industry aware of the R&D capabilities of laboratories.
- *Changing the performance evaluations and reward system for scientists.* Evaluation of the work of scientists and engineers should consider how well they market their research output. Scientists and engineers should share in the earnings from commercialization of R&D activities.
- *Upgrading facilities and investment in resource development.* Upgrading human resources is essential in an era of rapidly changing technologies, both to keep up-to-date with scientific advances and to train scientists in team building and client orientation.

Source: World Bank, 1997.

TECHNOLOGY EXTENSION CENTERS. To achieve international competitiveness, countries must not only generate new technologies, but also apply and diffuse them effectively. But potential users face uncertainty as well as information and learning costs, which may lead to underinvestment in technology. Suppliers of technology information and assistance also face learning costs and may lack experience in dealing with the structural barriers to promoting technologies. Most industrial countries have recognized these problems and have taken measures to promote technology diffusion, especially among small and medium-size enterprises. On the supply side are measures that seek to augment sources of information and assistance available to firms, and on the demand side are efforts to increase the willingness and absorptive power of firms to adopt technologies.

Shapira (1996) identifies key features of such programs and some examples:

- *Awareness-building and technology demonstration.* These measures seek to make potential users more knowledgeable about available technologies, their possible applications, and their benefits and costs. Japan's municipal technology centers (*kohsetsushi*) demonstrate new technologies to firms, often with hands-on training and pilot production. In the United States the Industrial Technology Institute of Michigan offers a performance benchmarking service, which allows companies to compare their use of technology with that of comparable and best-practice firms.

- *Information search and referral services.* These efforts aim to reduce information search costs, often matching user needs with appropriate resources. Denmark's Technological Information Centers, established in each county, offer information and other technical services to firms. For-profit companies, such as Teltech, Inc., in the U.S. state of Minnesota, offer specialized technology information services, matching corporate technology needs with appropriate sources of expertise. In several countries initiatives are under way to use the Internet for technical information exchange.

- *Technical assistance and consultancy.* These measures address limitations of expertise among both users and suppliers of technology, providing support to experts who can assess business problems, identify opportunities for upgrading technologies and industrial practices, and assist in implementing them. Technical assistance services are located in many applied technology centers. In the Valencia Institute of Small and Medium-Size Enterprise in Spain, trained staff offer technological advice and conduct assessments for firms in local industries. In some cases private consultants are engaged, through cost-sharing schemes, to assist particular firms.

- *Training*. Support for training (on the job, classroom, distance learning, management seminars, team-building workshops) addresses the tendency of technology users to underinvest in human capital, retarding deployment of new technologies and creating inefficiencies in their use once adopted. Measures to promote training for technology diffusion may also address deficiencies among institutions and vendors, which may need support to set up new technology courses. Australia has more than 60 Cooperative Research Centers that offer technology training focused on specific industry needs. Local Training and Enterprise Councils in the United Kingdom draw on public and private resources to identify and support appropriate training initiatives.

- *Collaborative research and technology projects*. Collaborative public-private research mechanisms address the gaps between technology development and deployment. They also seek to expedite commercialization of technological innovations and to focus research on key needs and opportunities. Applied technology centers around the world embody this collaborative research mode. In Baden-Württemburg, Germany, the quasi-public Steinbeis Foundation sponsors a system of some 130 technology transfer centers, often associated with polytechnic institutes, each conducting collaborative, industry-focused research. Japan's public technology institutes and new third-sector projects conduct applied research and technology projects with firms.

- *Personnel exchange and support of R&D personnel*. Small and medium-size enterprises, in particular, may lack the expertise to absorb new technologies or the resources to assign their staff to new research and technology projects. In Japan local public technology centers accept staff from smaller firms to participate in cooperative research and receive training in new technologies. In the United States the National Science Foundation and the Department of Commerce sponsor programs to place engineers in Japanese companies and research institutions. In Germany ministries have sponsored programs to subsidize research personnel in small and medium-size firms to help them absorb and develop new technologies.

- *Standardization*. Uncertainty about the compatibility of a technology can impede investments in its use. Diffusion can be accelerated by agreement between technology developers and users on standards and technological compatibility. In the United States the federally sponsored National

Information Initiative has promoted an industry-driven process of standards development. The development of standard measures for documenting quality, through ISO 9000 and subsequent reference marks, has also facilitated the diffusion of standard quality measurement techniques.

- *Financial support.* Instruments for reducing financial constraints to adopting new technologies include grants, loans and loan guarantees, interest write-downs, equity or near-equity investments, and royalty agreements. Public financial policies to promote technology diffusion often operate through intermediary institutions, including banks and quasi-government corporations. Examples include the preliminary cost sharing of private consultant assistance sponsored by the Minnesota Manufacturing Technology Center and grants through Italy's Act 696 to help small companies purchase high-technology equipment. The U.K. Support for Products under Research and the Small Firms Merit Award for Research and Technology programs support technology development in smaller firms, as does the U.S. Small Business Innovation Research program, which allocates a share of federal R&D budgets to support the development of small, technology-based firms.

- *Procurement.* Purchasing and specification policies by public institutions and large private firms can promote technology diffusion. U.S. defense procurement policies have favored small technology firms and the diffusion of new process technologies. Sometimes public support of investments in large firms is tied to conditions for local procurement, which may require supplier upgrading programs.

- *Interfirm cooperation.* Programs sponsor interfirm collaboration to promote technology diffusion by resolving common problems, sharing information, achieving scale economies in service provision and technology deployment, and strengthening business and technology development relationships. Collaborative efforts may be horizontal, among groups of small firms, or vertical, between suppliers and customers. In Finland applied technology and implementation programs have sponsored the formation of more than 200 collaborative groups of firms, both large and small. In Germany the Aachen Gesellschaft für Innovation und Technologietransfer helps groups of five or more companies identify common needs and develop collaborative R&D projects. In Japan's technology fusion program, groups of about 30 small companies work with local brokers and technology centers to commercialize new product technologies.

- *Regional or sectoral cluster measures.* In addition to building physical facilities, governments recognize the need to strengthen organizational capabilities and linkages within regions or industrial sectors. Institutional credibility and leadership, communication between technology developers and users and among users, and other aspects of social capital have proven extremely important in the diffusion of technology. Measures can involve strengthening industrial associations, promoting forums of stakeholders, building collaborative technology consortiums, encouraging labor-management collaboration, developing leadership strategies and shared visions, and strengthening links among users, service providers, and complementary public and private assets (such as banks or training institutions).

Linkage-Creation Institutions

Also important to the technology infrastructure of a country are institutions that link groups of firms, create synergies among firms and between firms and research institutions, and bring together startups and support services.

CLUSTERS. Studies are beginning to show that clusters of firms—firms linked through buyer-supplier relations, subcontracting relations, or formal or informal information exchange—tend to be more productive because of the information-rich atmosphere such clusters create (Schmitz and Nadvi 1999). The effect seems to be particularly strong for small and medium-size firms in both industrial and developing countries. The role of public policy in facilitating alliances among firms is not well understood, although local governments in Italy and Scotland have claimed success in facilitating cluster formation. Strengthening chambers of commerce and trade associations that supply members with information about a supplier or customer base, organizing trade fairs and study tours, and maintaining subcontracting databases and exchanges were some elements of the strategy in the Italian districts (Storper 1993).

TECHNOPARKS. Also known as technopoles, research or science parks, or science cities, technoparks focus on technology-intensive development, generally with a university or research institution as a key aspect of their establishment and operation. They are popular ways of building synergies among firms and between firms and universities or research institutions, which provide access to faculty, staff, students, libraries, laboratories, and technological infrastructure. Such operations are anchored by large, mature businesses that can make the substantial commitments required to take advantage of the technological potential of a nearby institution.

Interest in technoparks ("re-creating Silicon Valley" is a popular slogan) has surged for jump-starting economic growth in economies disproportionately concentrated on slow growth industries or hard hit by recessions. There was an explosion in technoparks in the United States in the late 1980s, for example. Nearly half of technoparks never reach viability, however, and nearly half of those that do are forced to diversify from research-based activities (Franco 1985).

Luger and Goldstein (1991) identify four factors underlying a technopark's success:

- Vintage, because it takes time to establish links with other businesses.
- Geographic region of the country, because that captures the local industrial base and political culture.
- Size of the population of the region, as a proxy for the presence of agglomeration and urbanization economies.
- Type of university, as a gauge of the bias toward research or teaching.

In addition, Luger and Goldstein identify an absence of effective leadership and commitment from top university officials as the primary cause of failure. They found that university officials often simply assumed that physically locating a research park close to a university would lead to close collaboration that would attract additional private R&D and boost innovative activity in the park. University leaders did not fully appreciate the difference in subculture between industries and universities and failed to provide sufficient incentives and institutional support to induce faculty to work with private industry. Incentives might include weighting collaborative work by faculty more heavily in the tenure and promotion process, providing salary supplements to encourage collaborative activity, and adding new faculty positions to departments willing to collaborate with industry.

INCUBATORS. Incubators house start-ups and provide them with basic services, ranging from reasonably priced real estate to financial, marketing, legal, and general business services. Firms occupy an incubator for three to five years on average, and during their stay they network with other tenants, engaging in joint problem solving, marketing, and business development. Technology incubators aim to help infant firms gain access to, apply, and market technical knowledge.

The case for public support of technology incubators usually rests on the argument that market or systemic failures impede the commercialization and diffusion of technology by new firms. The uncertainty associated with technology increases the risks for new business start-ups. Incubators, by reducing this uncertainty, increase the chances that new firms will survive. However, given the multiple objectives that incubators can be designed to pursue—creating jobs, increasing survival rates, boosting sales or profits, bringing new technologies to market—cost-benefit assessment is difficult.

Data on survival rates show an edge for incubator firms. According to one study (OECD 1997), in the United States, 80 percent of incubator graduates were still in business after five years, whereas only about 40 percent of firms in the general population still were. In Australia 50 percent of graduate firms were still in business after five years, compared with only 5 percent of other firms. And in France the failure rate was only 8–20 percent for incubator firms, depending on the technology group, but 31 percent for new firms in general. However, it is hard to say which way causality runs since intensive admission screening of firms may weed out likely failures, or self-selection might fill incubators with more dynamic companies seeking alliances with neighboring firms or universities.

Data on the impact of incubators on innovation uptake—the rationale for the existence of a specialized technology incubator rather than a general business incubator—have not been collected systematically. There is a general presumption that to have survived for very long, a technology firm must have engaged in innovative practices. Gauging by patents or technology licenses alone, tenant and nontenant firms showed no apparent difference in performance (OECD 1997). Tenant firms did have better links to higher education institutions, although they were no better informed about the research activities of the university. Thus questions remain about the cost-effectiveness of public support to incubators relative to alternative policy measures.

Lately, incubators have been set up to address the special needs of information technology firms that are seeking to develop or adapt information and communication technologies for conducting business (see Chapter 7 for a description of the investment experience). While it is too early to assess the economic viability of information technology incubators, it is reasonable to expect that the special support services (such as on-site banks in incubators serving as clearinghouses for e-commerce transactions) offered through incubators to commercialize these new technologies would do better in countries ranking high on overall e-readiness (see *http://www.infodev.org/* for an assessment of e-readiness methodologies).

Improving Access to Finance for Technology Development

While obtaining credit is a problem for firms in industrializing countries for many activities, finding sources of finance for technology upgrading is probably the hardest. Technology upgrading entails financing not only the purchase of new equipment, but also investments in adapting the equipment to local conditions and in learning how to integrate the new practices into the production process. These activities are often considered too risky and intangible for traditional lending instruments. Furthermore, long-term finance for firms is difficult to find in many developing countries, leading to a mismatch between the supply of funds, which are usually short term, and the demand for funds for technology development, which is intrinsically a long-term endeavor. Since the rate of return to investments in R&D is generally found to be several times that for investments in physical capital, this mismatch results in the loss of significant benefits to society.

LOAN AND GRANT ASSISTANCE. Several countries encourage investment in R&D activities by allowing tax write-offs. But only very large firms can afford to engage in formal R&D activities, and reported R&D activities are often exaggerated to take advantage of such tax benefits.

More successful has been direct financial assistance for technology upgrading, in the form of loans or grants, sometimes with matching finance (not necessarily 50–50) by firms. Matching finance secures the commitment of firms up front and minimizes the moral hazard problem associated with 100 percent outside funding. Matching loans can also be royalty based or made conditional on the success of the project, with a portion of the profits used to pay back the loan over a set period. Even though only firms that can put up a good portion of the project cost can qualify, a high degree of quality control is required to administer and manage conditional and royalty-based loans. A highly committed intermediary is therefore needed, and considerable technical assistance and training in the evaluation and selection of risky technology proposals are warranted (Box 5.4).

VENTURE CAPITAL. Venture capital supports high-risk investments in small technology-based firms, which are often passed over by large companies and traditional financial institutions. Venture capitalists invest in young companies that are not yet listed on the stock exchange, taking an equity stake. These investments are generally for three to seven years and involve a partnership with management to provide support and advice. Venture capitalists screen investment opportunities, structure the transaction, invest, and ultimately achieve a capital gain through the sale of their equity stake, through a stock market floatation, a trade sale, or a buy-back arrangement with the company.

Most venture capital schemes are independent funds that raise capital from financial institutions. Some financial institutions have their own venture capital funds (known as "captives" in Europe), and some countries have an informal venture capital market of private individuals ("business angels") and large companies ("corporate venturing").[1] Governments can play a role in fostering venture capital markets in a number of ways:

Box 5.4 Financial Assistance for Private Sector R&D: World Bank Experience in Turkey

Some $43.3 million of a $100 million loan to Turkey for a Technology Development Project, completed in 1999, was allocated for setting up the Technology Development Foundation of Turkey (TTGV). TTGV was intended to foster a technology culture in Turkey by providing seed capital to stimulate private investment in market-driven industrial technology development projects, on a matching finance basis.

Of the 273 applications TTGV received for R&D cofunding, 84 were selected. At the time of the World Bank implementation completion report on this project, 46 had been completed, with a total value of $99 million, $44 million of it from TTGV and $55 million from the private sector. About 64 percent of firms were small or medium size, and 36 percent were large. About 62 percent of the projects involved joint R&D with universities. Of the completed projects, 74 percent were technically and commercially successful, 20 percent were technically successful but commercially unsuccessful, and 6 percent were technically unsuccessful.

Commercially unsuccessful projects might nonetheless have added to the technical knowledge base of the company and induced firms to persist in R&D efforts. The successful projects resulted in a 5–45 percent increase in firm exports, depending on the year of completion.

Interviews with a sample of participating firms confirm the beneficial impact of the project. TTGV became well known and respected in Turkish industry. Several firms launched formal R&D programs for the first time, and others improved management of their R&D programs. TTGV activities also stimulated collaboration between industry and universities. Firms believed that the outside experts appointed as mentors of the funded projects made positive contributions to research and, more important, made firms aware of the potential benefits of interaction with technical experts from academia. Many of the research projects resulted in commercially viable products. Many were in technically sophisticated areas and helped to take the enterprise to the frontiers of technology and to match the best in Europe. In general, the TTGV program aroused considerable enthusiasm and interest in large-scale industry, earned a good reputation among industrialists, and established TTGV as a prestigious institution.

Source: World Bank, 1998.

- Promoting an investment culture by creating a fiscal and legal environment for stimulating the supply of venture capital, including measures to encourage longer-term venture capital investments by pension and

insurance funds and tax incentives for investments by individuals or businesses.
- Reducing risk for investors by stimulating the creation of venture capital funds dedicated to technology investments through appropriate tax incentives, seed financing schemes, guarantees for a proportion of investment losses, and funding of technology appraisals and audits before the financing is granted.
- Increasing liquidity by facilitating reinvestment and encouraging the creation of active secondary stock markets favoring high-growth technology-based companies and other means to ease exit by institutional investors in start-up ventures.
- Facilitating entrepreneurship by encouraging new high-technology start-ups through risk-bearing tax regimes, royalty-linked loan schemes, information and counseling services, and support for business angel networks.

Concluding Remarks

Fighting poverty through economic growth has long been a motivating principle of international development institutions. Supporting the private sector is increasingly seen as a way of supporting this goal, which in turn has inspired the support of privatization, legal and tax system overhaul, regulatory reform, infrastructure development, and similar programs. Strengthening the innovative capability of the private sector by encouraging technology development and diffusion is also fast becoming an important reform priority.

Much needs to be done in the newly industrializing countries to create or reform the infrastructure that firms need to be able to adapt and adopt new technologies. This chapter has suggested areas that could be strengthened and has discussed instruments of public policies for doing so. Not everything can be tackled at once, and the instruments must be chosen to match the market deficiencies being addressed. Only an assessment of the technology needs, capabilities, and support infrastructure in each country will reveal areas with the greatest room for improvement and that are likely to reap the highest value added from public intervention. Especially useful are tools for objectively assessing the innovation readiness of firms by evaluating the incentives facing firms and the capabilities of supporting institutions for reinforcing these incentives.

Note

1. Venture capital is well established in the United States, where it consists of a range of investors, including pension funds, insurance companies, and private individuals. The European venture capital industry is younger, oriented to main-

stream sectors, and dominated by banks. Japanese venture capital firms are mostly subsidiaries of financial institutions, which invest in established firms and provide mainly loan finance.

References

The word *processed* describes informally produced works that may not be commonly available through libraries.

Batra, Geeta. 2001. "Skills Upgrading, Technology and Productivity: Evidence from the WBES." Working Paper. World Bank, Private Sector Advisory Services, Washington, D.C. Processed.

Franco, Michael R. 1985. "Key Success Factors for University-Affiliated Research Parks." Ph.D. dissertation, University of Rochester, Rocheser, New York. Processed.

Henson and others. 2000. "Impact of Sanitary and Phytosanitary Measures on Developing Countries." University of Reading, Center for Food Economics Research, Reading, U.K. Processed.

Lall, Sanjaya. 2000. "The Technological Structure and Performance of Developing Country Manufactured Exports." Working Paper QEHWPS44. Oxford University, Queen Elizabeth House, Oxford, United Kingdom.

Luger, Michael I., and Harvey A. Goldstein. 1991. *Technology in the Garden: Research Parks and Regional Economic Development.* Chapel Hill, N.C.: University of North Carolina Press.

Maskus, Keith, John Wilson, and Tsunehiro Otsuki. 2000. "Quantifying the Impact of Technical Barriers to Trade: A Framework for Analysis." Working Paper. World Bank, Development Research Group, Washington, D.C. Processed.

OECD (Organisation for Economic Co-operation and Development). 1997. "Technology Incubators: Nurturing Small Firms." Report OECD/GD(97)202. Paris.

———. 1999. "An Assessment of the Costs for International Trade in Meeting Regulatory Requirements." Report TD/TC/WP(99) 8/FINAL. Paris.

Schmitz, Hubert, and Khalid Nadvi. 1999. "Clustering and Industrialization: An Introduction." *World Development* (U.K.) 27 (9): 1503–14.

Semerjian, Hratch, and Robert L. Watters, Jr. 1999. "Impact of Measurement and Standards Infrastucture on the National Economy and International Trade." National Institute of Standards and Technology, Gaithersburg, Md. Processed.

Shapira, Phillip. 1996. "Current Practices in the Evaluation of U.S. Industrial Modernization Programs." *Research Policy* 25: 185–214.

Storper, Michael. 1993. "Regional Worlds of Production: Learning and Innovation in the Technology Districts of France, Italy, and USA." *Regional Studies* 27: 433–55.

Tan, Hong, and Geeta Batra. 1997. "Technology and Firm Size-Wage Differentials in Colombia, Mexico, and Taiwan (China)." *World Bank Economic Review* 11: 59–83.

UNESCO (United Nations Economic, Social, and Cultural Organization). 1995. *1995 Statistical Yearbook.* Paris.

USITC (United States International Trade Commission). 1998. "Global Assessment of Standards Barriers to Trade in the Information Technology Industry." Publication 3141. Washington, D.C.

World Bank. 1997. "India: Investment Completion Report, Industrial Technology Development Project." Washington, D.C. Processed.

———. 1998. "Turkey: Implementation Completion Report, Technology Development Project." Washington, D.C. Processed.

———. 1999. "Turkey: Project Appraisal Document, Technology Development Loan II." Washington, D.C. Processed.

6
Upgrading Work Force Skills to Create High-Performing Firms

Geeta Batra and Hong Tan

Human capital contributes to economic growth by raising the productivity of workers and facilitating the adoption and use of new technologies. Support for this view is found in research on human capital and productivity, technology and innovation, and endogenous growth. In theory both education and training are thought to be important. In practice, however, studies have focused on the role of educational attainment, which is more readily measured than training.

Framework for Analyzing Enterprise Training

Evidence on the links between education, technology, and productivity is strong. Relatively little is known, however, about training and its effects on productivity, particularly in developing countries. Research based on answers to training questions in worker surveys finds that the likelihood of training and returns to training as reflected in wages are higher in industries where technological change is rapid, especially for the most educated workers (Lillard and Tan 1992; Tan and others 1992). However, because firm size and industry are often the only information available on firms in these surveys, little is known about the employer's role in training or about training's effects on firm-level productivity, which must be inferred indirectly from wages.

Training research using firm data is more limited. Bartel (1991) uses a sample of publicly traded U.S. firms to investigate the impact of training and finds that it has a positive effect on output, wage growth, and job performance. Tan and Batra (1995), using data from five developing economies, find that training has a positive and significant impact on firm-level productivity; Batra (1999, 2000) finds that formal training has a positive and significant impact on firm productivity in Guatemala and Nicaragua.

Despite these efforts, large gaps remain in our knowledge about training—its incidence among firms and in the work force, its determinants, and its consequences for firm-level productivity and economic growth. In many developing countries policymakers make critical decisions on resource allocation, and design education and training policies without reliable data on training. Often, the only data available are on the supply of graduates from public vocational-technical institutes and government training centers. Training policies thus tend to be supply oriented, and planning is often based on simple extrapolations of past trends in skill supply. A common policy response to perceived skill shortfalls is to expand the capacity of vocational-technical institutions. By failing to recognize that skill requirements change with shifts in demand, international competition, and technology, supply-oriented policies often result in mismatches between skills supplied by public training institutions and those needed by industry.

Why Focus on Firm-Level Training?

In most countries the largest share of training is provided by employers, either in-house or through external training institutions, equipment suppliers and buyers, industry groups, and joint-venture partners. When employers provide or sponsor training, matching training supply and demand is not an issue. Firms train only for needed skills. And because most new technologies enter developing countries through enterprises, employers have the equipment and technical information necessary to determine what skills are needed. Enterprises thus offer an important means to expand the resources available for skills development in the country.

Employers must decide not only whether to train, but whether to train in-house or to rely on outside training providers. In part, this decision depends on the ability of the vocational and technical education system in the country to meet the skill requirements of enterprises, the quality of technical training provided, and the job relevance of the skills that the training imparts. These factors determine how cost effective it is for enterprises to rely on outside training institutions rather than provide training in-house.

The technology literature suggests another set of determining factors. If the productivity advantage of new technology is revealed only through learning by doing, innovative firms have an incentive to train in-house to embody the new technology in its workers' skills. Outside providers are typically not well prepared to impart skills associated with the newest, still-evolving technologies. They play an increasingly important role when technologies become standardized and their productive characteristics well understood. In-house formal training has been shown to have the greatest impact on firm-level productivity, while informal on-the-job training has insignificant productivity impacts (Batra 1999, 2000; Tan and Batra 1995).

Data Sources and Sample

This chapter draws mainly on firm-level information collected through the World Business Environment Survey (WBES) in East Asia, which was administered to 502 enterprises in China, Indonesia, Malaysia, the Philippines, and Singapore in 2000 (Batra, Kaufmann, and Stone 2001), and on earlier work by Tan and Batra (1995) in Malaysia. It also provides comparators from the WBES for Latin America and member countries of the Organisation for Economic Co-operation and Development (OECD). The WBES includes information on firm size (employees, sales, and assets); years of operation; sales, debt, and growth performance; sources of finance; and a mix of qualitative and quantitative evaluations of the business environment, including corruption and governance, the regulatory regime, the predictability of economic policy, the nature of competition, public service delivery, the effectiveness of the judicial system, the availability of financing, and general constraints to business operations. For East Asia, OECD, and Latin America, the WBES data included a module on competitiveness, with rich information on enterprise training, technology, and productivity. All these data were used to estimate the relationships between training and firm-level productivity and to draw lessons for the design of training policy.

The WBES sample aimed to represent the relative importance of manufacturing and service and commercial firms in each economy. To ensure representative findings across countries, a sampling frame was developed for most of the regions, showing the distribution of privately owned companies in each country by sector, size, number of employees, and location.

Because the surveys oversampled larger firms relative to their weight in the population and results could not be weighted, aggregate training estimates should be viewed only as illustrative of broad patterns of training in countries. Comparisons by employee size, for which sampling weights are less of an issue, can be used to verify findings based on aggregate data. For this purpose, firms were divided into three size categories: small (50 or fewer employees); medium (50–499 employees); and large (more than 500 employees).

Who Is Trained—and How, Why, and with What Effect?

About 60 percent of surveyed firms provide some formal training. But the incidence of formal training reported by employers varies widely—from 65–75 percent or more in China, the Philippines, and Singapore to 30 percent in Malaysia (Table 6.1). On average, training increases with firm size. The exceptions are China and Malaysia, where a higher proportion of small and medium-size firms than of large firms report formal training. By sector a higher proportion of services firms provide training in Singapore and the Philippines, whereas a higher proportion of manufacturing firms provide it in China, Indonesia, and Malaysia.

A sizable proportion (40 percent) of firms in all five of the East Asian countries report providing no formal training—from 20 percent to 70 per-

Table 6.1 Percentage of Firms Providing Formal Training, by Country and by Firm Size and Sector

| | Firm size | | | | Firm sector | | |
Country	Overall	Small	Medium	Large	Manu-facturing	Services	Other
China	65.35	67	70	58	64.71	66.67	65.31
Indonesia	45.92	45	51	65	64.71	48.89	55.56
Malaysia	29	30	30	25	41.46	31.25	17.50
Singapore	76	79	69	78	72.41	83.33	73.17
Philippines	76.0	65	79	81	68.75	78.26	76.32
Total	60.48	57.73	59.89	65.32	59.85	64.52	57.84

Source: Batra, 2001.

cent of small firms and 20 percent to 75 percent of large firms. This is a worrisome finding considering the critical role of skills in technology development and the presumed beneficial effects of training on productivity growth.

Sources of Training

The survey distinguished between formal training provided in-house by the employer and formal training provided by external training institutions, both public and private (Figure 6.1). Reliance on internal training is strong across all countries, accounting for more than 40 percent of all training. Private providers account for more external training than do public providers.

Determinants of Formal Training and Productivity Outcomes

Earlier work in several developing countries found that the likelihood of firms investing in training is closely related to firm size, employee education and skills, investments in new technology, use of quality control methods of training, and foreign ownership (Tan and Batra 1995; World Bank 1997). The WBES survey tends to confirm those earlier findings. It also finds that small firms are much less likely to train than larger firms. Firms with better educated and more technically skilled employees are more likely to train, on the belief that such employees will benefit more from training. Investments in new technology, automated equipment, and quality control are also associated with increased training. Finally, local firms are less likely to train than are foreign firms, reflecting weaker training capabilities or lack of a training culture.

Figure 6.1 Sources of Training Provision, Selected East Asian Countries, Canada, and the United States

Percent of training

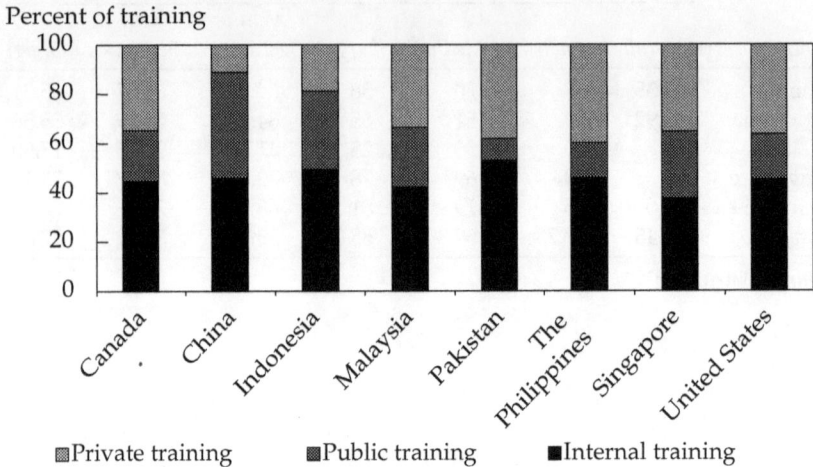

Recent research indicates that training increases firm-level productivity (Box 6.1). In Malaysia firms that train are 28 percent more productive than firms that do not (World Bank 1997). Training effects of this magnitude are not unusual for developing countries—some are considerably higher (Box 6.2). Productivity effects of training are higher in smaller firms, firms that are more export oriented, foreign firms, and firms using some form of technology (Figure 6.2).

A recent study using panel data for Malaysia found productivity growth to be highest with sustained training (World Bank 1997). For a balanced panel of firms observed in three years for which training data are available (1988, 1994, and 1996), the results show a strict ordering of productivity growth by number of years of training. Compared with the group that never trained in any of the three years, firms were 9 percent more productive if they trained in one year, 26 percent more if in two years, and 31 percent more if in three.

Two key findings emerge from this research. First, despite evidence that training increases productivity, a substantial fraction of firms, small and medium-size enterprises in particular, provide little or no structured training. Second, most of the productivity gains from training come from private training institutes and internal training programs. The limited training provided by public training institutions appears to have no significant impact on firm productivity.

Box 6.1 Productivity Effects of Training with and without Technology Investments, Taiwan (China), 1986

Using data from the 1986 Taiwan (China) Census of Manufacturers, Aw and Tan (1994) investigated the effects of training on firm-level productivity in seven industries, looking in particular at links with firms' technology levels, as measured by in-house R&D or purchases of technology. For each industry they estimated separate production functions for firms that invested in technology (high-tech firms) and for those that did not (low-tech firms), correcting for potential selectivity bias in firms' technology decisions.

Productivity Effects of Training—Taiwan 1986

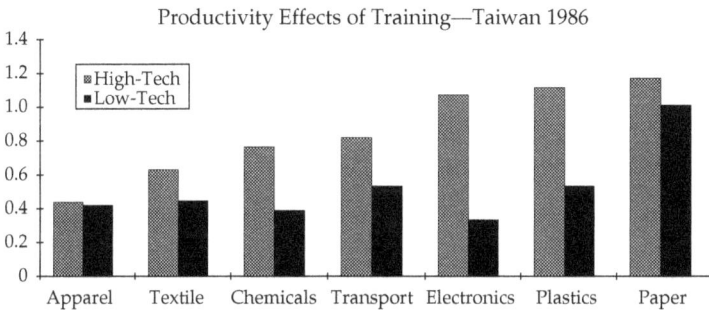

They found clear evidence that technology had an impact on the productivity outcomes of training. Within each industry training was associated with a larger impact on firm-level productivity when accompanied by investments in R&D or purchased technology. The differential impact of training in high-tech and low-tech firms was more pronounced in technology-intensive industries such as electronics, chemicals, and plastics than in more traditional industries such as textiles and apparel. Thus both within and across industries, returns to training were found to rise with technological change.

Source: Aw and Tan, 1994.

Constraints on Training from the Employer's Perspective

Considering the productivity gains from training, why do so few firms train? What constraints do they face in training? If market failures are responsible, what kinds of training policies effectively address these failures?

Do the low incidence of formal training and the striking differences in training by firm size reflect the weak training and technological capabilities,

Box 6.2 Enterprise Training and Productivity in Developing Countries

Using a simple production function approach, Tan and Batra (1995) and Batra (1999, 2000) estimated the productivity impacts of formal training. They found evidence that enterprise training is associated with higher firm-level productivity in all countries:

Country	Productivity effect (percent)
Indonesia (1992)	71.1
Nicaragua (2000)	56.4
Guatemala (1999)	49.0
Mexico (1992)	44.4
Malaysia (1994)	28.2
Colombia (1992)	26.6

The productivity effects of training are larger in lower-income economies (Colombia, Indonesia, Nicaragua) than in higher-income economies (Malaysia, Mexico), possibly reflecting the relative scarcity of skills in lower-income countries. No significant productivity effects were discernible for in-service training provided by public institutions.

or do they reflect market failures? Are firms constrained by poor access to financing for training, or do they lack the interest, know-how, or capability to design and implement training programs? How important a constraint is "labor poaching," the hiring away of employees trained at another employer's expense, which prevents firms from recouping the returns to their sunk investments in training? Answers to these questions are crucial for the design of public policy. Policymakers need to know what kinds of firms train and what kinds do not and why some firms invest little in training.

Insights into some of these questions were provided by East Asian firm respondents in the competitiveness module. Firm respondents were asked to rank on a scale of 1 (not important) to 5 (very important) the relevance of seven statements to their decision to provide little or no training:

- Training is not affordable because of limited resources.
- Training is costly because of high labor turnover.
- The firm lacks knowledge about training techniques and organization.
- The firm uses a mature technology, so learning by doing is sufficient.

Figure 6.2 Effects of Training by Firm Type in Malaysia, 1994

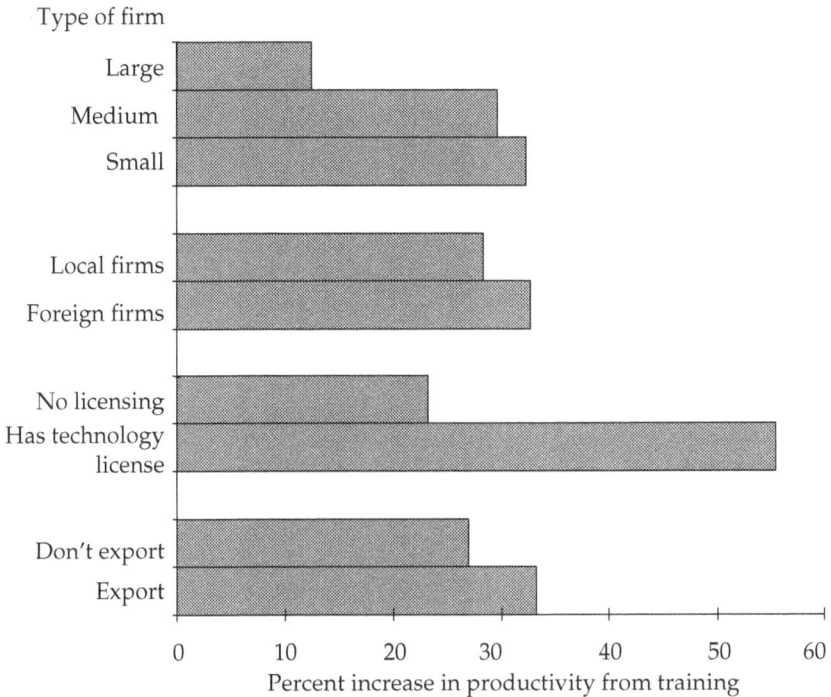

Source: World Bank, 1997.

- Informal training is adequate.
- Skilled workers are readily hired from other firms.
- We are skeptical about the benefits of training.

For the overall sample the use of mature technologies was the most commonly cited reason for providing little or no training (45 percent), followed by the belief that informal training is adequate (35 percent). The productive attributes of mature technologies are well established, and there is typically little scope for improving on existing production techniques. Workers trained on the job quickly become proficient, so informal training is adequate. And when mature technologies are widely diffused, the supply of skilled labor with experience using the technologies is generally plentiful. The third most important reason for not training is that high labor turnover makes training costly by preventing employers from recouping their investments in workers' skills (33 percent). These rankings are consistent by firm size as well (Figure 6.3). However, small firms are more constrained by resources, lack of knowledge about training benefits, and labor turnover than are large firms.

These findings are not specific to East Asia. In a recent study on Latin America, manufacturing firms cited use of mature technologies, ready availability of skilled labor and high labor turnover among the important reasons for little or no training (Figure 6.3)(Batra 2001). Thus there is evidence for East Asia and Latin America that not enough firms provide training and that market failures are important constraints for many employers, especially small and medium-size enterprises.

Policy Prescriptions and Experience Promoting Training Markets

Many countries, both industrial and developing, have put into place various policies designed to increase in-service training among its enterprises,

Figure 6.3 Why East Asian Firms Provide Little or No Training, by Firm Size

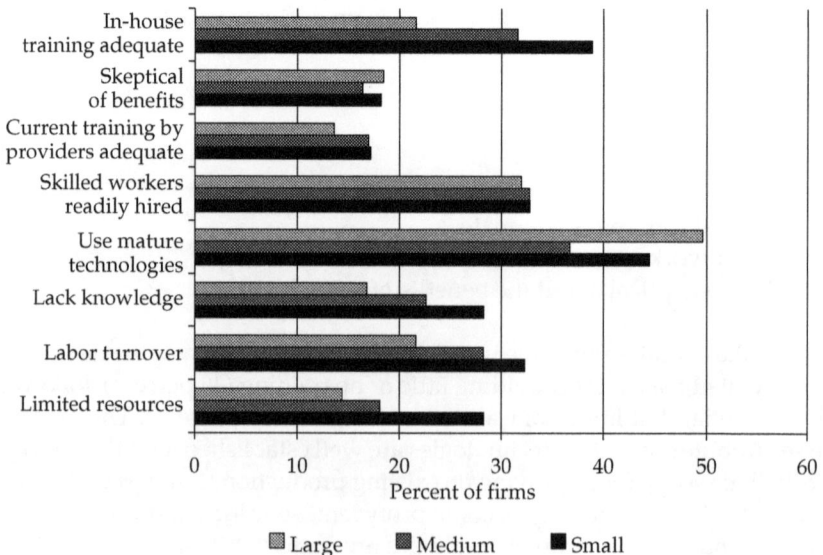

Percent of firms

☐ Large   ◼ Medium   ◼ Small

Note: Reasons were categorized as important if firms assigned them a rating of 3, 4, or 5 on a scale of 1 (not important) to 5 (very important).
Source: Batra, Kaufmann, and Stone, 2001.

including payroll-levy training funds and tax incentives for employer-sponsored training, national training councils, and matching grants. This section focuses on some of these public policy instruments employed in various countries and provides some best-practice examples.

The kind of policy intervention that is appropriate depends on the nature of the market failure. When poor information is the constraint, for example, the appropriate policy response is to disseminate best practices in training know-how and information about the availability and cost of services. High rates of labor turnover suggest that there are externalities in training. To the extent that firms are unable to internalize the benefits of training because skilled workers can be hired by other firms, there will be underinvestment in training. The appropriate policy response is to require all firms to train or to contribute to the cost of training provided by others in the industry.

The Payroll Levy Grant System

The payroll levy is a common instrument in Latin America for overcoming the underprovision of training. Argentina, Barbados, Brazil, Chile, Colombia, Costa Rica, Ecuador, Honduras, Jamaica, Paraguay, Peru, and Venezuela have implemented levies, with mixed success. The ratio of enrollments in training and vocational education courses to enrollments in secondary education in Latin America is more than double that for other developing country regions, with more students enrolled in training centers than in vocational schools. Brazil's levy, one of the oldest, suggests that factors influencing success include administrative independence of the levy fund, a combination of private ownership and public mission, and a management structure that includes industry and government.

In Chile small and medium-size enterprises have grouped together on a sectoral basis and thus overcome the tendency for levy funds to oversubsidize large firms. Sectoral centers have also been established in Argentina, and small and medium-size enterprises have negotiated a sectoral rate with the national supervisory agency.

Levies can also be set and controlled through industry sector bodies (Mexico, South Africa), industry collective agreements (Belgium, Denmark, France, Greece, the Netherlands, Sweden), or national insurance levies for displaced workers (France, Sweden). Levy exemption schemes have also been used as a means of subsidizing smaller firms and training providers (Austria, Germany). Similar levy schemes have been implemented in Botswana, the Democratic Republic of Congo, Morocco, and Turkey. Levy-rebate schemes have been used in Botswana, Côte d'Ivoire, Jordan, the Republic of Korea, Malaysia, Mauritius, Nigeria, Singapore, South Africa, Taiwan (China), Tunisia, and Zimbabwe. Some countries have modified

---

Box 6.3 Well-Designed Levy-Grant Schemes Can Motivate Firms to Train

Several East Asian economies have used direct reimbursement of approved training expenses, funded through payroll levies, to encourage firms to train their employees. Successful schemes are flexible, demand-driven, and often accompanied by an information campaign and technical assistance to smaller firms. The introduction of such a scheme in Taiwan (China) led to dramatic increases in training, which continued after the program ended in the 1970s. Singapore uses a levy on the wages of unskilled workers to upgrade worker skills through the Skills Development Fund. The fund's aggressive efforts to raise awareness of training among firms, to support development of company training plans, and to provide assistance through industry associations have led to a steady rise in training, especially among smaller firms. However, such schemes can also create disincentives to train when rigidly administered. In Korea, which required that training last a minimum of six months or that firms pay a fine, many firms paid the penalty rather than train to this standard.

Source: Tan, 2000; World Bank, 1997.

---

payroll levies. Peru lowered the levy and diversified the sources of finance. Argentina, Brazil, and Colombia have cofinancing arrangements with enterprises, communities, and vocational schools. Nigeria, Singapore, Taiwan (China), and Zimbabwe have used matching grants.

Experience with levies yields several lessons:

- Industry needs to own the levy. Argentina, Brazil, and Chile have vested supervision of levies in industry bodies.
- Levy funds are not cost effective when they support only government training providers.
- Funding levels are more readily maintained with levies than with government grants, which have tended to decline with shrinking government budgets.
- Levy exemption or reimbursement schemes have typically been used mainly by large enterprises and enterprises that already have a high skills base. It is important to establish support and advisory mechanisms for small and medium-size enterprises to participate in these programs.
- Care must be taken to avoid using levies for government financing (as in Costa Rica).

An example of best practice is the Human Resource Development Fund (HRDF) in Malaysia, which was established in 1993 with a matching grant from the government. (The government contributed R48.9 million to match projected company levies in the first year; in each of the following three years, it will add an additional R16.3 million to the HRDF.) It replaced the training tax incentive scheme (the double deduction incentive for training), which had been in operation since 1987 and which was widely acknowledged to have been relatively ineffective. The HRDF created a council (HRDC), with representatives from the private sector and from responsible government agencies, and a secretariat to administer the HRDF schemes. Eligible employers with 50 employees and above are required to contribute 1 percent of payroll to the HRDF. Those who have contributed for a minimum of six months are then eligible to claim a portion of allowable training expenditures up to the limit of their total levy payments for any given year. The HRDC set rates of reimbursement, varying by type of training; reimbursement was generally lower for larger firms (Tan 2000).

National Training Councils and Levy Administration

Many countries have established national training councils (Argentina, Australia, Brazil, Chile, Côte d'Ivoire, Malawi, Mauritius, South Africa, United Kingdom). Their experience suggests that locating country management of training with the social partners (business, unions, and government) can improve the quality, relevance, and flexibility of training. Training funds managed by training partners have tended to become more diversified in their sources and uses, including their use in the informal sector.

In Argentina, Brazil, Chile, Mauritius, and Peru, industry associations have assumed responsibility for administering the levy. It has been important for these industry bodies to have responsibility for the bulk of the funds and to work with both training providers and enterprises.

The Japan Industrial and Vocational Training Association is an association of employers that provides training programs for industry trainers. The association receives no funding, but charges membership and course fees. The semi-autonomous Vocational Training Corporation in Jordan is an industry body that works closely with government and industry in providing in-house and external training.

In the United Kingdom, Training and Enterprise Councils, industry associations that administered training funds at the regional level, are being replaced by regional Learning and Skills Councils that will combine a broader range of education and training functions.

There are many mechanisms for distributing training funds. Financing can go to state-run training institutions, it can be directed selectively to enterprises on the basis of training plans (Germany, Korea, Singapore), or it can be distributed through open tender, with the state as purchaser rather than supplier of training (Australia, Chile). A more radical measure distributes funds to the user or trainee through voucher schemes; this is the approach taken under the United Kingdom Training Credits scheme.

Because different types of firms and workers require different types of training, it is important that the training market not be too bound by institutional constraints. In Colombia, despite an extensive and well-funded training system, many skilled workers have not used the formal training system.

Matching Grants Schemes

Some countries use matching grants schemes to increase training. The most successful schemes are demand-driven, with private sector implementation, and aim to facilitate the creation of sustained markets for training services. Programs in Chile and Mauritius use private sector agents to administer the matching grants. Both have reported positive results.

So has Mexico's program. An increased investment in training has been matched by a reduction in enterprise failure. A side benefit has been the development of a network of industry management training consultants who are available to enterprises that want to invest in enterprise-based training (Box 6.4). Singapore has undertaken a program to build up its stock of industry trainers, and Japan's Industrial and Vocational Training Association has trained more than 30,000 industry trainers in the past 30 years.

Matching grant schemes can support the development of a training culture by providing both an incentive for and a means of investing in training. It is important to build a training culture with a high level of training capacity in enterprises and a high propensity for workers to undertake training, so that enterprises continue to invest in training. In Japan most managers have a training function and regularly engage workers in informal training. The Basic Law for Vocational Training in Korea is designed to encourage in-company training. Strong training cultures have been established in some Asian countries (Japan, Korea, Singapore), some northern European countries (Germany, Netherlands, Scandinavia), and, judging on the basis of levels of in-company training, some Latin American countries (Brazil, Chile).

Matching grants schemes can also link educational and human resource development policies. The Singapore Skills Development Fund was designed and successively modified to provide an incentive for enterprises to increase the skill and pay level of their workers.

A matching grants scheme, by itself, will not lead to an expansion of the training market. Grants should not be restricted to state-run training institutions. Funds should support strengthening and diversifying the supply

Box 6.4   Mexico's Proactive Approach to Small and Medium-Size Enterprise Support

The Integral Quality and Modernization Program (Programa de Calidad Integral y Modernización [CIMO]), established in 1988 by the Mexican Secretariat of Labor, has proven effective in reaching small and medium-size enterprises and assisting them to upgrade worker skills, improve quality, and raise productivity. Set up initially as a pilot project to provide subsidized training to small and medium-size enterprises, CIMO quickly evolved when it became apparent that lack of training was only one of many factors contributing to low productivity. By 2000 CIMO was providing an integrated package of training and industrial extension services to more than 80,000 small and medium-size enterprises each year and training up to 200,000 employees. Private sector interest has grown, and more than 300 business associations now participate in CIMO, up from 72 in 1988.

All states and the Federal District of Mexico have at least one CIMO unit, each staffed by three or four promoters; most units are housed in business associations that contribute office and support infrastructure. The promoters organize workshops on training and technical assistance services, identify potential local and regional training suppliers and consulting agents, and actively seek out small and medium-size enterprises to deliver assistance on a cost-sharing basis. They work with interested enterprises to conduct an initial diagnostic evaluation of the firm, as the basis for training programs and other consulting assistance. CIMO is expanding its support in two directions: assisting groups of small and medium-size enterprises along specific sectoral needs; and providing an integrated package of services, including information on technology, new production processes, quality control techniques, and marketing as well as subsidized training.

Evaluation found CIMO to be a cost-effective way of assisting small and medium-size enterprises (Tan 2000). This study tracked two groups of small and medium-size enterprises over three years, one with firms that participated in CIMO in 1991 or 1992, another with a broadly comparable control group of enterprises that had not participated in the CIMO program. CIMO firms tended to have lower performance indicators than the control group before participation in the program, but by 1993 labor productivity in these firms had either caught up to or exceeded that of the control group. Other performance indicators showed similar improvements—increased profitability, sales, capacity utilization rates, wage and employment growth, and reduced labor turnover, absenteeism, and rejection rates for products. The most dramatic impacts of CIMO interventions were among and small firms.

Source: STPS, 1995; Tan, 2000.

of training and stimulating demand. Mexico's Integral Quality and Modernization Program concentrates on the productivity of the enterprises, using both private training consultants and government and private training institutions.

Training Programs for the Informal Sector

Training for the informal sector is typically provided as informal apprenticeships, often through nongovernmental organizations, which help to diversify the funding for training programs to the poorer sections of the economy. Argentina, Costa Rica, and Peru have successful programs of this type.

Informal sector activities can vary from traditional rural and agricultural work to urban manufacturing enterprises. Apprenticeship training is relatively common, but the circumstances of workers vary considerably. Workers may be school leavers or displaced workers, rural or urban, literate or illiterate. The relationships between informal and formal sector enterprises vary as well. Consequently, the nature of training programs for the informal sector has varied. Programs may concentrate on basic literacy, traditional occupations, or parallel trades. The Botswana Brigade, for example, provides small rural-based training centers for school leavers. Successful training programs in the informal sector concentrate on the productive process and the problems encountered. An example is a literacy program established in Tibet through a program targeted at pest control in agriculture.

Concluding Remarks

Productivity analyses in developing countries find that investments in training, especially in-house training, have large payoffs. Yet a sizable fraction of manufacturing and services firms do not provide any formal training for their employees. This is especially pronounced for smaller firms—over half of them provide no structured formal training. A significant number of large firms also report no training. The evidence indicates that several constraints on training—poor information about benefits, high training costs from the inability to exploit scale economies in training, weak managerial capabilities, absence of competitive pressures, or market imperfections—may be operative and that policy initiatives to address these constraints should be explored. Firms that train use a variety of in-house and external providers. Private sector providers are as important as—if not more important than—government-run training institutions.

Firms are more likely to train when they are large, employ an educated and skilled work force, invest in R&D and technology, emphasize quality control methods, have foreign capital participation, and export to foreign markets. Thus there are strong complementarities between training and schooling and critical links between firms' training, technology, and exports.

These findings have implications for training institutions, training policy, and small and medium-size enterprise policies. *To be effective, development policies should reflect this interdependence of human resource and industrial strategies.* Important recommendations include:

- Better collection and dissemination of training information to monitor and analyze private sector training efforts and the efficacy of public policies in promoting worker training and productivity growth.
- Expanded role for education and training institutions. In-service training is a key part of human resource development. Employers' decisions to train and the productivity outcomes of training depend on the stock of education and technical skills that individuals bring to the labor market.
- More effective training policies. Training policies need to be well designed and implemented, and targeted firms need to be made aware of the policies.
- Better coordinated and proactive policies designed to encourage small and medium-size enterprises to train. A high percentage of small and medium-size enterprises do not train. Such firms face a variety of training constraints, from high labor turnover to poor information and finance. Proactive measures are needed to seek out and deliver a package of integrated services to small and medium-size enterprises—including consultancies, training, and technology information and incentives.

References

Aw, Bee Yan, and Hong Tan. 1994. "Training, Technology, and Firm-Level Productivity in Taiwanese Manuacturing." World Bank, Private Sector Development Department Working paper. World Bank, Washington, D.C.

Bartel, Ann. 1991. *Productivity Gains from the Implementation of Employee Training Programs.* NBER Working Paper 3893. Cambridge, Mass.: National Bureau of Economic Research.

Batra, Geeta. 1999. "Skills Upgrading and Competitiveness in Guatemala." World Bank, Private Sector Advisory Services, Washington, D.C. Processed.

————. 2000. "Private Sector Training and Competitiveness in Nicaragua." World Bank, Private Sector Advisory Services, Washington, D.C. Processed.

————. 2001. "Skills Upgrading, Technology and Productivity: Evidence from the WBES." Working Paper. World Bank, Private Sector Advisory Services, Washington, D.C. Processed.

Batra, Geeta, Daniel Kaufmann, and Andrew Stone. 2001. "Voices of Firms: Findings from the World Business Environment Survey." World Bank and International Finance Corporation, Foreign Investment Advisory Services, Washington, D.C.

Lillard, Lee, and Hong Tan. 1992. "Private Sector Training: Who Gets It and Why." In Ron Ehrenberg, ed. *Research in Labor Economics*, vol. 13.

STPS (México Department of Labor and Social Welfare). 1995. *La Capacitacion y Asistencia Tecnica en la Micro, Pequena y Mediana Empresa: Evaluacion del programa CIMO.* México City.

Tan, Hong. 2000. "Do Training Levies Work? Malaysia's HRDF and Its Effects on Training and Firm-Level Productivity." World Bank Institute Working Paper. World Bank, Washington, D.C. Processed.

Tan, Hong, and Geeta Batra. 1995. *Enterprise Training in Developing Countries.* Washington, D.C.: World Bank.

Tan, Hong, Bruce Chapman, Chris Peterson, and Allison Booth. 1992. "Youth Training in the U.S., Great Britain, and Australia." In Ron Ehrenberg, ed., *Research in Labor Economics*, vol. 13. Grennwich, Conn.: JAI Press.

World Bank. 1997. "Enterprise Training, Technology and Productivity in Malaysia."

7
Reaping Efficiency Gains through E-Commerce

Prasad Gopalan

Electronic commerce—better known as e-commerce—has come to occupy center stage in economic discussions because of its potentially far-reaching economic and social benefits. E-commerce shrinks the distance between consumers and producers by eliminating several layers of retailers and distributors. Although some new services will be needed (network access providers, for example), overall e-commerce promises a substantial reduction in intermediation links, which would reduce transaction costs, lower entry barriers, and improve the quality of information about products for consumers, thereby promoting competitiveness.

There are several definitions of e-commerce, but no single definition captures its full potential. The discussion in this chapter takes a firm-level perspective and defines e-commerce as the use of information technology to conduct business transactions among buyers, sellers, and trading partners to improve customer service, reduce costs, and increase shareholder value. The chapter looks first at the claimed benefits of e-commerce and projections of demand for e-commerce activity. Then to illustrate the policy issues involved in setting up an e-commerce facility, the chapter presents a case study of an e-commerce project in the Philippines.

Benefits of E-Commerce

The chief benefits claimed for e-commerce fall under cost savings and productivity gains.

Reduction in Costs

E-commerce has the potential to lower costs associated with executing a sale, procuring production inputs, making and delivering products or services, and managing logistics. One of the biggest effects on the cost of executing a sale comes from the replacement of bricks and mortar infrastructure with virtual storefronts. Instead of several separate physical establishments, one e-commerce website can be accessible to a global market round the clock. Cost savings from processing payments over the Web are also sig-nificant. The administrative marginal cost to a bank for processing a paper check averages $1.20, and merchants pay an average of $0.50 for debit or credit card purchases, whereas the cost of processing an electronic pay-ment over the Internet can be as low as $0.01 or less. Finally, more efficient order configuration also yields cost benefits. Both GE and Cisco Systems reported that nearly a quarter of their pre-Web orders had to be reworked because of errors. Since adopting a Web-enabled customer interface, Cisco reports an error rate of only 2 percent.

The value of Web-based procurement of maintenance, repair, and oper-ations supplies was around $100 billion in 2000. (Maintenance, repair, and operations supplies are the goods required to run a company, as distinct from the raw materials used in the first manufacture of a product or pro-vision of a service. They are generally low-cost items purchased intermit-tently in bulk.) Under a traditional procurement system, a purchasing officer responds to a paper-based requisition by searching a variety of paper cat-alogs to find the right product and price. Often, the administrative cost exceeds the unit value of the product itself. The Organisation for Economic Co-operation and Development estimates that companies with more than $500 million in revenue spend $75 to $150 to process a single purchase order for maintenance, repair, and operations supplies. By electronically linking organizations to preapproved suppliers' catalogs and processing the entire purchase on the Web, e-commerce applications have reduced costs dra-matically. In fact, large firms such as Ford insist that their suppliers link into their Web-based procurement systems as a condition of doing busi-ness with them.

Rather than increasing production and inventory in advance of customer demand, businesses are striving to make their supply chains respond in real time to actual sales. Electronic supply chain management makes the entire chain visible by allowing any or all portions to be tracked in detail. Alert messages are communicated to all concerned parties when prede-fined tolerances are violated—for example, if a shipment is more than five hours late. Such messages are aggregated and analyzed to help firms iden-tify bottlenecks, assign responsibilities between carriers and suppliers, and better understand their own distribution costs. Some analysts point out that an important implication of e-commerce is that it involves a shift from competition between Firm X and Firm Y toward competition between the supply chain of Firm X and the supply chain of Firm Y.

E-commerce transforms logistics from the simple packaging and moving of goods to an information business by tying carriers (such as Federal Express) closer to product shippers and their customers through electronic load ten-dering, inventory confirmation, and delivery tracking. As more businesses move to build-to-order process models and extremely low inventory levels, a higher value is placed on prompt, accurate inbound and outbound logistics. Even companies that traditionally did not worry about tracking parcels have expressed interest in switching to Web-based applications to reduce customer service calls, monitor carriers, announce delays, and provide delivery verification, all of which help to control costs. Companies also expect to reduce costs significantly by increasing flexibility in production schedules while reducing unnecessary transportation and warehousing expenses.

Contribution to Productivity

Recent evidence from the United States shows a substantial contribution of information technology (of which the Internet is a component) to productivity growth (Baily and Lawrence 2001). There was a 1.8 percentage point structural increase in labor productivity in the private nonfarm business sector between 1973–95 and 1995–2000. Of this, 0.6 percentage point came from capital deepening in computer communication and software capital, 0.2 from total factor productivity growth in the computer hardware industry, and 1.0 from faster total factor productivity growth in goods and services other than computer hardware.

How much of the 1.0 percentage point growth in productivity can be attributed directly to information technology? Evidence at the industry level shows that much of the acceleration in labor productivity took place in the services (non-goods-producing) sector. Productivity growth increased much faster in the large wholesale and retail trade, finance, and business services than in the rest of the economy. These industries are heavy users of information technology products that help improve supply chain management (nearly 70 percent of information technology products are purchased by wholesale and retail trade, finance, and telecommunications firms). Evidence also shows that more intensive users (based on information technology spending) within the services sector experienced much faster productivity growth than others.

Projected Demand for E-Commerce

Between 1995–97 and 2000–02, forecasters estimate (median value of 12 estimates) that e-commerce activity will soar from $725 million to $154.5 billion, a 200-fold increase. The number of users who make purchases over the Web is projected to jump from 31 million in 1998 to more than 183 million in 2003. Business-to-business activity dominates, at 90 percent of the activity, with business-to-consumer activity accounting for the remaining 10 percent.

Although the number of Web users is increasing worldwide, Internet commerce is currently centered in the United States. In 1998, 56 percent of Web users resided outside the United States, but non-U.S. Internet commerce accounted for only 26 percent of worldwide spending. International Data Corporation estimates that by 2003, 65 percent of Web users will be international and that countries other than the United States will account for slightly less than half of world Internet commerce. In the Asia Pacific region (excluding Japan), as the digital divide closes, the number of users making purchases on the Web is expected to increase from 1.2 million at the end of 1998 to 12.7 million by the end of 2002.

There is also reason to believe that e-commerce will boost international trade substantially. Even though the telephone and facsimile made speedier communication available decades ago, only e-commerce allows businesses to attract the equivalent of "walk-in" traffic from another country. The ease with which users in one country can access a Web site to compare, negotiate, and contract over the Internet is unparalleled.

The ePlanters Business Concept and Rationale: A Case Study on Investing in E-Commerce

Using e-commerce to leverage technology, thereby increasing operational efficiency and broadening market reach, is still uncommon among many small and medium-size enterprises in developing countries. The International Finance Corporation (IFC) recognized the need to provide such enterprises with low cost and quick access to the Internet, e-commerce, and advisory services. The ePlanters project in the Philippines aimed to meet the market demand for such services.

Two Service Lines of the ePlanters Business Model

The IFC, along with the local sponsor (Planters Development Bank) and the technical partner (Vicor), created a Philippine company to provide small and medium-size enterprises with Internet business services to supplement their conventional product placement and distribution channels and provide customized business support. The services of ePlanters would help smaller enterprises leverage the competitive advantages of e-commerce, such as ease of transaction, extended market reach, and increased operational efficiencies.

Before the ePlanters project, the proposed business model and certain variations had been introduced with some success in serving small businesses in more developed markets (mainly the United States) by Vicor and others such as Yahoo! (Yahoo! shops), Intel (iCAT), and Amazon.com (zShops). The ePlanters business model incorporates two service lines to meet the market expansion and global market information needs of small and medium-size enterprises: an electronic storefront, and a resource center to serve small and medium-size enterprises.

The Web-based business model involves development of computer applications including a menu of eStore designs; a step-by-step process to create an eStore; and ordering, settlement, and delivery options. The storefront can be expanded with additional application services, such as electronic banking, payroll services, finance management, accounting services, and a variety of other operational business devices. For a reasonable one-time cost and monthly service fee, which includes setting up a storefront in a matter of hours, enterprises are able to receive orders and payments over the Internet, reducing transaction costs, increasing sales volume, and improving profitability. These electronic transactions allow clients to leverage technology to achieve operational efficiencies.

The ePlanters business model transcends traditional online merchandise exchanges, adding value by providing resources on industry trends and business education. The resource center, a portal for small and medium-size enterprises, collects and collates industry and business information from the Internet and provides proprietary tools, such as on-line document templates, customer profiles and business credit checking, and financial analysis, all adapted to meet the local business environment and knowledge needs of small and medium-size enterprises. The resource center provides quick and easy access to key information for growing client businesses that would otherwise be difficult and expensive to obtain.

The Asian Institute of Management is scheduled to manage the resource center's content, which is expected to include free links to on-line information on business planning, financial management, tax planning, and the like; e-commerce advisories; news and reports; links to other small and medium-size enterprise information sites; chat rooms and bulletin boards, "ask the experts," and frequently asked questions.

Partnerships

The commercial viability of ePlanters depends on strategic alliances and partnerships. The business model requires a local project sponsor with established local relationships, brand recognition, and a network for marketing ePlanters' services. The IFC teamed with Planters Development Bank (Plantersbank) as sponsor for the project, with Vicor for technical management and the Asian Institute of Management for managing resource center content.

The International Finance Corporation, a member of the World Bank Group, was established in 1956 to encourage private sector activity in developing countries. It does so primarily by financing private sector projects, helping companies in the developing world mobilize financing in international financial markets, and providing advice and technical assistance to businesses and governments. The IFC offers a full array of financial products and services to companies in its developing country members:

- Long-term loans in major currencies, at fixed or variable rates.
- Equity investments.
- Quasi-equity instruments (subordinated loans, preferred stock, income notes).
- Guarantees and standby financing.
- Risk management (intermediation of currency and interest rate swaps, provision of hedging facilities).

The IFC can also help structure financial packages, coordinating financing from foreign and local banks and companies and export credit agencies. To be eligible for IFC financing, projects must be profitable for investors, benefit the economy of the host country, and comply with stringent environmental guidelines. The IFC finances projects in all types of industries, from agribusiness to mining, from manufacturing to tourism. Although it is primarily a financier of private sector projects, the IFC may provide finance for a company with some government ownership, provided there is private sector participation and the venture is run on a commercial basis. It can finance companies that are wholly locally owned as well as joint ventures between foreign and local shareholders.

To ensure the participation of investors and lenders from the private sector, the IFC limits the total amount of debt and equity financing it will provide for any single project to 25 percent of total estimated project costs; it may provide up to 35 percent of the equity capital for a project provided it is not the largest shareholder. IFC investments typically range from $1 million to $100 million. Its funds may be used for permanent working capital or for foreign or local expenditures in any World Bank member country to acquire fixed assets.

Plantersbank, the primary sponsor of ePlanters, is a private development bank officially organized as a thrift bank. With total assets equivalent to about $420 million at the end of June 1999, Plantersbank is the largest bank in the Philippines specializing in lending to small and medium-size enterprises. It is the 28th largest bank in the Philippines in total assets.

The IFC's partnership with Plantersbank gives ePlanters a key competitive advantage over future competitors. Plantersbank's extensive portfolio of small and medium-size enterprise clients includes more than 2,000 borrowers and 40,000 depositors in diverse sectors. Plantersbank is targeting some 50,000 new customers through the PlantersClub membership campaign. Plantersbank has an extensive network with more than 50 branch offices and business centers throughout the Philippines. This network is the optimal channel through which to market ePlanters' services.

Plantersbank is also positioned to leverage its own growth from ePlanters, as the bank currently has e-banking operations under development. Through ePlanters, Plantersbank will also be able to provide e-commerce services to its customers, giving it a competitive advantage over other banks.

Vicor, founded in 1989, is a privately held e-business providing an array of Internet-based e-commerce products and consulting services to help clients gain strategic advantage by applying advanced technologies. The company is employee- and partner-owned, with no outside debt or owners. Among Vicor's clients are major commercial and investment banks, securities trading companies, insurance companies, healthcare businesses, and small businesses.

Vicor's core competencies include Web and needs-based publishing; complex data, document imaging, and video capture; high-volume transaction processing; large-scale storage and retrieval systems; payment systems connectivity and settlement; and Internet multimedia networks. Vicor has introduced the proposed e-business model in the United States in conjunction with Bank of America (through bankofamerica.com/merchantservices), in the United Kingdom (Safestreet.com), and in Canada (electronictradingpost.com).

The Asian Institute of Management (AIM) is the leading business school in Southeast Asia, with connections to the dot.com community. Through the Center for Entrepreneurship, established in 1996, AIM has focused on developing entrepreneurs in the small and medium-size enterprise sector. With its keen interest in promoting Internet-based distance learning programs for this sector, AIM was a natural choice for providing content for ePlanter's resource center. Furthermore, AIM was willing to provide experts to seed ePlanters' chat room, while AIM's Center for Entrepreneurship offered a captive test audience for piloting ePlanters concepts.

Due Diligence

Several due diligence issues needed to be resolved before the project could go forward. These included confirming a market demand for ePlanters services, deriving appropriate pricing of the services, developing a shareholding structure, determining necessary capitalization, and obtaining necessary approvals and authorizations.

The ePlanters partners assessed market demand through a top-down study of the Philippine small and medium-size enterprise sector. The target market was Plantersbank's small and medium-size enterprise clients with businesses suitable for e-commerce and with an interest in outsourcing their e-commerce capabilities. The general market included firms that had similar characteristics but that were not Plantersbank customers. The commercial viability of ePlanters was predicated on the ability to penetrate these two markets.

Of approximately 35,000 small and medium-size enterprises in the Philippines, 23,000 (65 percent) were estimated to have businesses suitable for e-commerce. The ePlanters project team estimated that the project would be unprofitable until the third year of operations, so capitalization would need to be substantial enough to fund early year losses. The total project cost was estimated at $1 million (Table 7.1), with a proposed venture capitalization shareholding structure, as shown in Table 7.2.

In April 1999 the IFC, Vicor, and Plantersbank held focus group discussions to reconfirm the market demand for ePlanters and the pricing scheme. The sessions confirmed the positive assessments of the prospects for such a company and for the proposed fee structure.

Table 7.1 Estimated Project Cost, ePlanters

Cost	Amount (U.S. dollars)	Share (percent)
Start-up costs	340,000	34
Software customization	200,000	
Hardware	60,000	
Premarketing	40,000	
Leased line (T-1)	40,000	
Working capital (funding for two years)	600,000	60
Salaries and benefits	360,000	
General and administrative costs	40,000	
Marketing	180,000	
Miscellaneous expenses	20,000	
Contingencies	60,000	6
Total	1,000,000	100

Source: Company data, Planters Development Bank, IFC, and Vicor.

Table 7.2 Proposed and Actual Venture Capitalization Shareholding Structure

Equity holder	Amount (U.S. dollars)	Proposed share (percent)	Actual share (percent)
Plantersbank	500,000	50	55
International Finance Corporation	250,000	25	
Vicor	200,000	20	
Asian Institute of Management	50,000	5	5
Total	1,000,000	100	100

Source: Company data, Planters Development Bank, IFC, and Vicor.

Revenue

Revenues for ePlanters are generated from transaction fees, Web hosting fees, and advertising. Clients pay a one-time set-up fee of $100 and can add customizations for additional charges. The estimated total average set-up fee is $150 per client. Clients can choose from among a number of Web page templates designed by Vicor and upload their product catalogs into the selected template. Clients are charged a $35 monthly Web-hosting fee at the ePlanters website (www.plantersbanke-commerce.com) and a 1 percent transaction fee on all Web-based transactions. Because e-banking is currently illegal in the Philippines, transactions are settled offline. Advertising space is also sold on ePlanters' site.

Technical Issues

Lack of affordable and reliable access to telecommunication links is an important impediment to the adoption of e-commerce in developing countries. For example in China, despite rapid economic growth and a large population, inferior infrastructure has constrained the adoption of e-commerce.

Governments can help to improve the pricing, choice, and quality of telecommunication services by adopting an effective policy and regulatory framework that delineates the appropriate roles of the public and private sectors in telecommunication service delivery and addresses aspects of digital convergence of telephone, cable, and wireless technologies. Another key role for the government would be to develop and implement an effective competition policy for the information technology sector (see Chapter 3). Finally, given the dynamism of information technology, industry associations should be encouraged to develop standards for service levels and quality in the Internet service provision sector.

Lack of access to computing facilities is another impediment to the adoption of e-commerce in most developing countries. This is primarily an affordability issue and is related to the price of computers, software, and ancillary equipment. Accession to the Information Technology Agreement under the auspices of the World Trade Organization should result in the elimination of tariffs and duties on semiconductors, computers, and ancillary devices, making them more affordable.

Finally, e-commerce implementation is only as good as the technology supporting the underlying business. E-commerce allows Web shoppers (businesses or consumers) to buy products and services safely and efficiently over the Internet. Vicor created the hosting environment, which houses the operating system and Web server software that manages the information flow on www.plantersbanke-commerce.com. This includes site creation software, catalog shopping cart/storefront software, payment processing modules, and tax and shipping charge calculations. For these software and hosting services, Vicor charged ePlanters a Web design and set-up fee of $310,000 and quarterly servicing fees of $45,000. At any time ePlanters has the option to purchase the programming code from Vicor for $900,000.

Development Impact

With small and medium-size enterprises constituting more than 90 percent of the Philippine business community, the development and economic growth of the Philippines would benefit from these firms' adopting e-commerce solutions, with their associated competitive edge. In that context the IFC's role in ePlanters was to:

- Foster cross-border partnerships to implement Internet-based solutions for small and medium-size enterprises in the Philippines.
- Provide seed capital for a start-up company.
- Bridge diverse interests and play an honest-broker role.
- Implement the project concept through active board participation.
- Learn, and replicate the experience in other markets.

The three target developmental roles of Plantersbank were to:

- Assist the formation of a pioneering company in the Philippines focused on helping small and medium-size enterprises by mobilizing appropriate local partners with the ability to reach the targeted enterprises.
- Identify an appropriate technical partner that could transport cutting-edge business concepts and technological solutions to small and medium-size enterprises in the Philippines.
- Invest in a start-up company and mobilize additional seed capital for the project company.

The developmental role of ePlanters for small and medium-size enterprises was to:

- Provide access to resources and business concepts to enable small and medium-size enterprises to grow.
- Enable small and medium-size enterprises to sell products and services in a global marketplace through a new Internet channel.
- Serve as a model for Internet-based e-commerce solutions for small and medium-size enterprises in other World Bank Group client countries.
- Expand small and medium-size enterprise product placement channels at an affordable cost.

- Provide small and medium-size enterprises with quick and easy access to valuable tools and services to grow their business.
- Enable its banking partner to deliver value-added customer service by participating in the e-business revolution.
- Enable small and medium-size enterprises to increase productivity and efficiency.

The ePlanters project also fits the World Bank and IFC strategies for small and medium-size enterprise development and the World Bank's Development Gateway initiative by facilitating technological leapfrogging for companies that wish to participate and helping them compete in a global marketplace. The ePlanters project supports the IFC's small and medium-size enterprise strategy for the Philippines by:

- Improving access for small and medium-size enterprises to tools and business concepts through the on-line resource center.
- Allowing small and medium-size enterprises instantaneous access to the Internet and helping position their products in a global marketplace.
- Providing seed capital for an Internet start-up focused on small and medium-size enterprise businesses.
- Enabling the project sponsor, Plantersbank, to gain a deeper understanding of the needs of the small and medium-size enterprise sector and to better serve them.

Main Issues

Three sets of policy issues relating to ePlanters emerged over corporate governance concerns, the Vicor technical service agreement, and regulatory and policy considerations. Corporate governance of ePlanters includes a five-member board of directors consisting of two representatives from Plantersbank and one each from the IFC, Vicor, and AIM. Minority rights are protected through antidilution provisions, including preemptive rights and supermajority requirements on key board decisions (a minimum of 80 percent of shares voted are needed to pass key resolutions). Minority investors also have tag-along rights (the right to sell shares on the same terms as the sponsors).

Because ePlanters lacks a direct link to the small and medium-size enterprise community independent of Plantersbank's branch network, a key corporate governance issue is to ensure Plantersbank's commitment to the project. Therefore, Plantersbank was required to retain a minimum of 35 percent ownership in ePlanters, and a lock-in period was placed on the Plantersbank's and Vicor's equity subscriptions. Finally, to avoid possible conflicts of interest, all related-party transactions are prohibited, unless on an arm's length basis.

The use of Vicor's proprietary technology (the Web-hosting environment and associated applications to operate ePlanters) and Vicor's position as the system administrator magnified the importance of the technical services agreement. Vicor is on a technical support service retainer and also hosts the ePlanters server in California with a backup presence in the Philippines. Under this retainer ePlanters is entitled to software and hardware support, free upgrades, and training and support for two years, with options for renewal. The technical service agreement also covers the software license/right of use. The license is for five years with options for renewal. Under this arrangement ePlanters is entitled to add-ons developed on a nonexclusive basis, with the software becoming the property of ePlanters. ePlanters has an option to buy the source code at any time for $900,000 (although it is barred from reselling the code).

Under National Telecom Commission regulations, ePlanters can have maximum foreign shareholding of 40 percent. Under the proposed shareholding structure, the IFC and Vicor together would have had a 45 percent shareholding. An ePlanters petition to the NTC for an exemption was denied, so the IFC and Vicor could subscribe to only 39.9 percent of ePlanters. To keep capitalization at $1 million, Plantersbank subscribed to an additional 5 percent.

Other necessary authorizations included clearance by the Philippines Central Bank for financial institutions, registration of the technical service/license agreement with the Intellectual Property Office, and arrangements pertaining to security and privacy, terms and conditions for end users and merchants, and online authorizations.

Concluding Remarks

The advent of e-commerce also raises other important issues that must be resolved. These include taxation, privacy, protection of intellectual property, harmonization of national laws regarding such commerce, and strengthening human capital, among other things.

Taxation of e-commerce is a challenge. Paperless transaction and settlement capabilities, the disappearance of convenient taxing points due to disintermediation, and the blurring of international boundaries place pressure on tax authorities. Nonetheless, certain underlying principles may be used to guide taxation of e-commerce: tax rules should be fair, simple, and sufficiently flexible to keep abreast of technological progress and commercial developments. Above all, they should introduce as little economic distortion as possible.

Privacy and security issues drive consumer confidence in conducting transactions over the Web. Consumers need to be assured that their communications (including on-line payment instructions with credit cards) and data are safe from interception and unauthorized use and that appropriate mechanisms are available for redressing violations. Public policies need to focus on ensuring that transactions over electronic networks are safe (using firewalls and encryption technology) and verifiable (public "key" infrastructure that makes encryption work). Protecting consumers against fraud could involve enacting legislation against computer crimes, amending and enforcing consumer protection acts (to protect against erroneous transactions), and providing certain lawful access to encrypted information.

E-commerce often involves the sale and licensing of intellectual property. Without sufficient protection of the terms and conditions under which content providers and application service providers place their work on public domains, e-commerce offerings might not be expected to realize their full potential. Participation of countries in the World Intellectual Property Organization as well as enactment and enforcement of stringent intellectual property protection codes should provide the requisite confidence and lower the risks for e-commerce service providers in making their work available. Logistical issues involving affordable, on-time delivery and the ability to return faulty goods are also important to customer satisfaction and continued use of e-commerce services.

In an increasingly global economy e-commerce offers an ideal medium for cross-border transactions. But this, in turn, requires harmonization of legal and regulatory provisions across countries. Among barriers to the growth of e-commerce in many developing countries are inefficient customs procedures and the inadequacy of legal and regulatory frameworks for the formation and enforcement of contracts.

Contract laws were developed in a paper-based world, and most legal instruments require written, signed, or original documents. Electronic documents create unique problems for proving who wrote a document and verifying the exact contents of an agreement, partly due to the ability to change data stored on computers. Perfection of on-line contracts and admission of electronic data, including electronic signatures, as evidence in courts are still in question in many emerging market countries—posing an undue risk for both consumers and merchants to conduct e-commerce. One way to avoid inconsistencies in commercial codes across countries is to base e-commerce laws on the United Nations Commission on International Trade Law model law on e-commerce. Policymakers need to focus on removing uncertainties in contracting and on making coordinated decisions across their own national boundaries to encourage e-commerce.

Human capital development, through education and training, drives people's understanding and, ultimately, adoption of e-commerce. Generation of a high degree of local e-commerce activity is unlikely without widespread, basic computer literacy. In addition, several transactional issues warrant consideration. First, convenient payment and settlement systems must be devised, with both in-country and international modes of settlements. Second, effective e-commerce business models must be developed that emphasize the transfer of knowledge from e-commerce-enabled economies (like the United States) to developing countries. Finally, financing matters, including prudent company laws, commercial codes, financial and securities regulations, and a well-functioning capital market, play a major role in attracting venture capital to companies facilitating e-commerce.

References

Baily, Martin Neil, and Robert Z. Lawrence. 2001. *"Do We Have a New E-conomy?"* Working Paper 8243. National Bureau of Economic Research, Cambridge, Mass.

PriceWaterhouseCoopers. *Technology Forecast: 2000.* PriceWaterhouse-Coopers Technology Center, Menlo Park, Calif.

8
Value Creation through Supply Chain Management

Ronald Kopicki

A paradigm shift in industrial organization is taking place around the world. Business processes are being reengineered, reduced to their essentials, and linked one to another through globe-spanning telecommunications systems. Suppliers of components, original equipment manufacturers, and distributors are connecting and networking their operations in unprecedented ways. The effects of these changes on the ways companies compete with one another are as fundamental as those that resulted from the introduction of assembly line manufacturing techniques and vertical corporate integration in the first and second decades of the 20th century.

This shift has accelerated the pace and sharpened the precision of basic business transactions. It has compressed business processes in time and extended them across space, effectively coordinating production scheduling, shipping, and inventory control with final sales on a worldwide basis. This change in basic business processes has far-reaching implications for economic development and for the public policies that affect national competitiveness.

Supply Chains and New Forms of Industrial Organization

The new paradigm extends process integration, coordination, and control beyond individual fabrication plants both upstream and downstream into all links in the value chain. Multiple industrial, distribution, and consumption activities can be managed across organizational boundary lines. Standardized controls and process integration capabilities now available from third parties encompass the entire set of sourcing, producing, delivering, consuming, and even recycling activities. The entire value chain can be integrated on a worldwide basis and actively managed through the Internet.

Just as mass production techniques and new business models jarred complacent industrialists in the first and second decades of the 20th century, competition based on supply chain methods and associated new business models is beginning to jar traditional competition. Competition is based not simply on new process technology or even on the reduced cost of production or superior marketing strategy alone but, more fundamentally, on new forms of industrial organization. Increasingly, companies compete with one another not only in one-on-one market combat, but also for affiliation with superior commercial partners, as parts of larger commercial systems. They compete through the efficiencies and the adaptability of the supply chains of which they are a part.

In this business environment, transactions become more than arm's-length exchanges of value. Facilitation of low-cost transactions, the transparency that requires, and the kinds of affiliation and trust entailed by transparency of total supply chain inventory and real time transfers of market and competitive product information take on strategic significance. In their formative stages transactions among supply chain partners consist of exchanges of information, cash flow, and risk, as well as of goods sold from downstream vendors to upstream buyers. Supply chains can be thought of as conduits for the flow of goods, cash payments, information, and ownership rights among trading partners.

Supply Chains and Knowledge Sharing

As supply chains mature, however, they become systems for efficient knowledge transfer. They facilitate the sharing and application of knowledge among participants—including knowledge about customer preferences, competitive pricing, competitive product design, and process technology. Over the long term it is knowledge sharing that enables some supply chains to maintain their competitive advantage over other supply chains. The old adage that a chain is only as strong as its weakest link applies with ruthless precision in the new information age. Hence, maintaining market, product, and process currency becomes the responsibility of each member of the chain.

Supply chains can operate across borders and in diverse cultural and national environments because uniform commercial standards now make it possible not only to move products efficiently across borders, but also to integrate supply chain processes across borders. Supply chain partnering is made possible by connectivity and interoperability among processing networks, standardized resource planning systems provided and maintained by third-party vendors, independent third-party exchanges for transport services, cooperative forecasting of production based on sophisticated real-time Bayesian methods, third-party information partnerships, third-party cash management networks, and commercial applications software for linking optical scanners at checkout counters to plant-based order fulfillment systems. Harmonized standards for trade financing, transport documentation, carrier custodial responsibilities, and even standardized customs clearance protocols ensure that risks associated with doing business across international links in supply chains are minimized, that specific risks can be assigned to the supply chain partners best qualified to manage them, and that widely dispersed production and distribution platforms are interconnected.

Increasingly, transactions are being entered into preferentially among supply chain partners that not only adopt compatible commercial and process standards, but also integrate their internal process management and control systems. As part of their operational coordination agreements, supply chain partners sometimes even share fixed assets and operate them jointly. They exchange information openly and adopt appropriate file structure standards to facilitate these transparent exchanges. They also redesign their internal work activities and assume mutually supportive work assignments to minimize redundancy within the chain. The results are accelerated cash flow and cost and time savings, which are available to be shared among supply chain partners. Specialized third-party service providers offer services that leverage the synergies realized through the cooperative commitments of supply chain partners. Increasingly, the tactical management of supply chains is being outsourced to third parties.

The kinds of cost-saving and market-enhancing synergies that once could only be realized through mergers and consolidations are now being realized between supply chain partners bound only by a shared strategic vision or a common competitive challenge and by joint marketing or operating agreements that explicitly anticipate contingencies and responsibilities—and remedies for each.

Examples of Supply Chain Adaptability

Some of the best documented successes in the application of advanced supply chain management methods are in the consumer products industry, where "continuous replenishment" of vendor-managed inventory was developed through a partnership between Procter & Gamble and Wal-Mart. This continuous replenishment system has become a best-practice standard that has been broadly imitated and improved by several "category buster" retail chains in North America. In reengineering inventory management relationships with their own core vendors, surpassing the Procter & Gamble–Wal-Mart standard has become their objective.

Other examples come from the personal computer industry. Dell Computer has refined its Internet-based retail order fulfillment process and progressively compressed its order-production-retail delivery time so that its response time is measured and managed in hours rather than days. Similarly, Intel has progressively refined its flexible manufacturing methods so that it no longer appears to be constrained by traditional unit production cost–response time tradeoffs. Intel's supply chain cost structure varies with volume over a broad range of fixed asset utilization levels. Cisco Systems has developed backward information links to its contract manufacturers and forward links to its customers that allow it to operate with a precision and dexterity that set new standards of excellence.

In the roiling stock market for high technology of 2001, a "failure to execute" within previously untested parameters of falling market demand was the single most important test of company value among financial analysts and investors. These stock value tests, applied with harsh effect to many high-tech firms in 2001, are in reality supply chain tests.

In the automotive industry Toyota sets the pace for supply chain excellence based on its continuous improvement of time-to-market performance. Toyota relies more extensively on vendors for component subassemblies than do any of its international competitors. Despite this dependence and its worldwide original equipment manufacturing network, Toyota has succeeded in compressing design-to-manufacture, procurement-to-production, and order-to-delivery response times of its vendor-dependent network sufficiently to redefine the industry supply chain paradigm from make-to-forecast to make-to-customer orders. This shift has given Toyota an edge not only on the supply side (lowest unit cost, for example), but increasingly on the demand side as well.

Another interesting supply chain innovation under way in the automobile industry is being tested by Volkswagen in its newest production facility in Brazil. Volkswagen has challenged its vendors not only to manage inbound logistics more efficiently, but also to apply subassemblies to frames, as part of a vendor-executed final vehicle fabrication plan.

In all three of these industries—consumer products, computers, and autos—institutional walls that once separated firms and frequently led to adversarial relations are becoming both transparent and mutually supportive. Competitors that cannot develop comparably flexible and value-creating relationships with their own vendors and customers fall further behind industry leaders.

Competition is becoming more based in the supply chain in other industries as well. Xerox has developed a winning strategy based on superior after-sales service, equipment buybacks, and remanufacturing of office equipment from recycled parts. Xerox operates a "reverse logistics" network of installers, parts reclaimers, and disassemblers as a franchised network of third parties. Through this network Xerox is able to reclaim component parts and subassemblies, breaking them down into reusable components and shipping them directly from its regional network of recycling centers to its worldwide network of equipment fabrication plants. Xerox designs its component subassemblies to hold up to multiple cycles of use. Its new equipment similarly is designed around core functions and components that are interchangeable with previous generations of equipment. Xerox has given life-cycle cost minimization a new significance. The resulting cost savings and customer price flexibility (in allowing variable discounts for recycled equipment based on the opportunity cost of the reclaimed parts they contain) have given Xerox a substantial edge.

In the search for core competencies to sustain their competitive advantage in the face of new challenges—including rapidly changing customer preferences, shorter-lived process technologies, more frequent product launches, and more numerous market entrants—these companies have learned that only the ability to adapt supply chain responses rapidly and flexibly can be relied on to sustain enhanced shareholder value over the long run. Supply chain management has become the core competency for many successful corporations (Figure 8.1).

Macro Effects of Supply Chain Development

The efficiency with which supply chains operate and the flexibility with which they can reconfigure themselves in response to new competition are primary sources of national competitive advantage. Supply chains do not spontaneously emerge in all policy environments. In an important sense they are quasi-public goods, requiring vision and planning, supportive preconditions, and clarity of rules and regulation from the public sector.

Figure 8.1 Supply Chain Improvement Is a Sustainable Process When It Is Driven by Competition

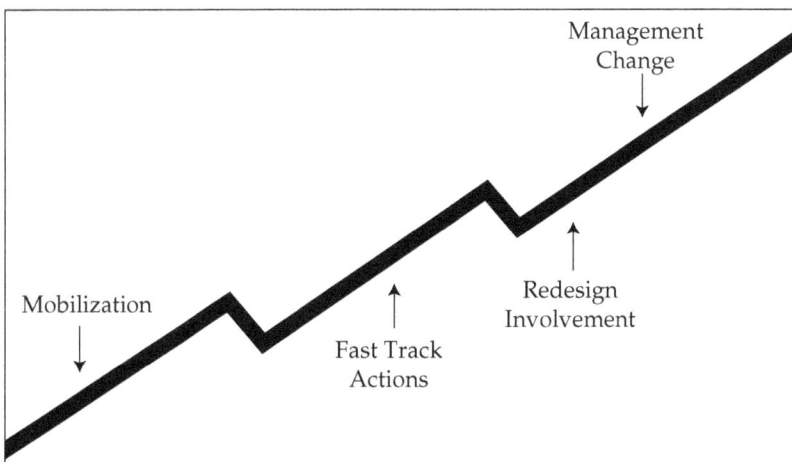

Without public sector involvement, supply chain capabilities will be under-invested. An economy that makes too few supply chain investments will not be able to increase its growth potential as fast as an economy in which supply chain capabilities are fully invested and resulting competitiveness benefits fully realized.

The logistics attributes of a national economy—for example, the efficiency with which buy-sell-deliver transactions are completed—determines in large measure the share of globalization benefits an economy can enjoy. Logistics efficiency and compatibility of the global trade network are as important in determining the attractiveness of an economy for foreign direct investment at the beginning of the twenty-first century as sovereign risk was in the twentieth century. Thus, for example, the ratio of aggregate transactions (or, said another way, aggregate logistics cost) to gross domestic product (GDP) determines how efficiently working capital turns over within an economy and how efficiently this application of capital, which represents 70 percent or more of capital investment in emerging market economies, performs.

This ratio also determines the productive return that can be realized on investment in fixed plant and equipment. This has both a substitution aspect and a turbo-charging aspect. Capital committed to complete buy-sell-deliver transactions is not available for investment in productive fixed assets and human capital. Thus the difference in development potential between a lean, high-service economy whose aggregate-logistics-cost-to-GDP ratio is 10 percent (like the United Kingdom or the United States) and a working capital–intensive, low-service economy whose aggregate-logistics-cost-to-GDP ratio is 30 percent (like many developing countries) is significant. In the high-service economy 20 percent of GDP is free for productive reinvestment. In the low-service economy 70 or more additional days are required to clear distribution channels before new products or processes can be introduced. In the turbo-charging aspect, investment in supply chain capabilities leverages the productivity of investment in fixed assets and human capital because it accelerates the turnover of working capital that these assets produce. Hence economies with low levels of aggregate logistics cost are more attractive targets for direct foreign investment.

At the macro level supply chain efficiency significantly determines how well internal markets clear and how quickly supply responses adjust to changes in demand. Logistics costs can be thought of as a buffer between aggregate demand and aggregate supply. Every economy has some minimum efficient buffer or aggregate logistics cost, just as it has a minimum employment level. Both are determined by the level of technology implemented within the country and the level of investment in market-clearing capacities to improve the turnover of working capital, such as more efficient customs systems, less government-imposed transaction cost and process delay, faster cash transfers within the banking sector, and lower cost resolution of claims and contract liabilities.

The challenge of maintaining competitiveness in the global economy is intensifying progressively for developing economies. As the service expectations of prospective customers in industrial countries are continuously ratcheted up, so too are their supply chain service expectations. Global competitiveness in the 21st century is about more than selling low-cost, high-quality products. Increasingly, it is about the quality and reliability of the service package in which these goods are delivered to customers. It is about the ability of sellers to integrate their outbound logistics and order fulfillment systems with the inventory control systems and inbound logistics systems of prospective customers.

Factors Affecting Supply Chain Development

What has driven this paradigm shift? The simple answer is fundamental changes in the global business environment and new technology. The more elaborate answer involves several key factors, including information technology, the emergence of time-based competition, techniques for synchronizing production and inventory with final sales volume, the outsourcing of value-adding functions, globalization, improved methods of supply chain management, and supportive public policy.

Information Technology

Supply chain development is enabled by information technologies. These modern technologies provide the information network structures that link elements of the supply chain. In the early 1970s remote terminals (with modems) were able to access central computers for the first time through publicly switched telephone networks. In the mid-1970s distributed computer processing began to emerge, and distributed computer networks grew into wide area networks. Again, these were supported by publicly switched digital networks. As the technology advanced, local area networks appeared using client-server structures and private telecommunications networks. Both types of networks operated primarily within a single organization. During the 1980s a new third-party service industry began to offer network users a variety of specialized digital services over public networks. Value added networks provided electronic bulletin boards, e-mail, and a wide variety of other community software products. In the 1990s the low cost and service coverage features of the Internet and its open TCP/IP protocol provided an ideal medium for lacing together global supply chains.

Today investment in information technology is being traded off for investment in inventory and in process cost. Thus, for example, in traditional supply chains real-time information about retail sales and inventory levels was shared throughout the chain via electronic data interchange or dedicated private networks. Sharing real-time information allowed supply chain partners to reduce their individual costs of holding inventory without increasing the risk of lost sales.

With electronic commerce, opportunities to compress supply chains and to accelerate free cash flow have multiplied severalfold. For example, with business-to-consumer commerce, orders can be taken and filled from vendor inventories and cash payments received even before accounts payable are booked. Competitor-cooperators can scan each other's inventories and deliver products to their customers from inventories that are closer to the customer, saving time and transport costs.

Sharing information also allows process redundancies to be minimized. Thus, for example, data entry, reworking, and physical inspection of received goods or processed subcomponents can be reduced, along with transaction costs, to levels that allow new kinds of security interests to be taken in inventories in storage, transit, or delivery on consignment. So-called supply chain bonds are emerging as a new kind of asset-backed security. This new source of financing requires controls and systems that did not exist before the information age.

Information technology also enables supply chain partners to plan and schedule work collaboratively. With the assistance of community decision support software, real-time distributed decisionmaking can be supported over an entire network. A new mode of networked management has arisen that is decentralized, collectivist, and inclusive. With the assistance of structured decision support software, new kinds of networked organizational forms can be structured with only the limitations implicit in the software.

Information technology also enables the flexible assignment of risks among supply chain participants. For example, transaction support software makes it possible to convey contingent ownership and control rights to inventories irrespective of their physical location. It also allows for contingency contracts among supply chain partners that, in lieu of less flexible merged corporate structures, can distribute risks among chain participants and third-party insurers in ways that reduce the net cost of otherwise uncovered risk.

Finally, information technology simplifies the search for new supply chain partners. Commercial information systems facilitate the recruitment and operational engagement of uniquely well-qualified supply chain partners who can satisfy the precise standards of low cost, commercial competence, and technical capability required of legacy partners. Markets can be created outside the chain for new value-adding competencies, and new kinds of virtual supply chains can be formed.

Information technology has made the integration of geographically dispersed, value-adding activities into globe-spanning supply chains so routine that long-standing tradeoffs between "making" and "buying" have now shifted radically in favor of buying. Thus information technology has opened new strategic opportunities for outsourcing, partnering, and reinventing buyer-vendor relationships. It has opened the possibility for implementing hitherto unknown business models.

Time-Based Competition

In the earlier period of information scarcity, the fundamental management challenge was to discover and analyze adequate information to allow better decisionmaking. In the current period of information sufficiency (some would say information surplus), the fundamental challenge facing management is to transform information into action.

Time-based competition is all about achieving competitive advantage through reduced organizational response time—for example, by designing new products faster than the competition, bringing new products to market faster, or reducing order-to-delivery time or production-to-delivery time.

Time-based competition is also about using agility strategically to compensate for a lack of information about future developments and contingencies. In a business environment where change is the only certainty and the pace of change appears to be accelerating, rapid response becomes a valuable survival skill. Agile and adaptable firms are able to change direction faster than their competitors even when they cannot anticipate market turns any faster than their competitors.

Time-based competition is also about using assets more productively. It is about investing directly only in high-velocity (high-turnover) assets and relying on chain partners' capital (trade credits, leasing, or short-term borrowing) to provide lower-velocity assets. Asset utilization and asset productivity rates have a great deal to do with balancing production and distribution capacities among chain elements and with coordinating process technology and other productivity-enhancing investments in a larger chain context.

In general, however, faster is better in the world of supply chains. Improvements in a supply chain system can ensure ever faster inventory turnover, faster cash-to-cash cycle times, shorter time spent in transport and shorter payback periods on all investments in the chain. Box 8.1 shows some of the time-based parameters of greatest concern to supply chain managers.

Techniques for Synchronizing Production and Inventory with Anticipated Demand

Supply chains are interconnected, with end-to-end process controls. Every activity in the chain is dependent on every other activity in terms of activ-

---

Box 8.1 Time-Based Parameters Relevant to Supply Chain Design

The time-based parameters of greatest concern to managers include:

- Average level of working capital committed/cash flow cycle.
- Average working capital/average sales per day.
- Time from order to delivery.
- Time from product concept to manufacturing prototype.
- Time from procurement to fabrication.
- Inventory/average sales per day.
- Return on working capital.
- Working capital turnover.
- Customer response time.
- Competitive product response time.
- Production cycle time.
- Number of days of inventory buffer.

---

ity or process rates, inventory accumulation, and shipment frequency. Controls within the chain ensure that the performance of each participant is scripted to balance the entire system and that each activity is executed precisely according to a chainwide plan.

Elements in most highly efficient supply chains operate in an anticipatory rather than a response mode. This makes the system highly interdependent and reliant on flawless execution by each link. Quality control techniques ensure that the need for reworking or for special handling is minimized along the chain. "Routine" process steps are typically less expensive than "special" process steps. Execution failures are kept to a minimum, but when they occur the system must be ready with a rapid recovery response.

Typically, information about demand for products at the retail end of the chain is captured at the point of sale and shared with all members of the chain. Additional efforts are made to ensure that inventory stocks are transparent and visible to all supply chain partners. Supplies are typically pulled rather than pushed through efficient chains. This arrangement allows for whipsaw action within the chain, with corresponding buildup and runout of inventory at multiple order fulfillment points. Under the best of circumstances, retail inventories are replenished continuously just as products are sold.

Continuous efforts are made to compress both order-to-delivery and production-to-storage times and to push the reorder point forward in the supply chain so that it becomes more responsive to final demand. Efforts may also be made to postpone investment in working capital until later in the chain and to maximize product–packaging–target market flexibility until late in the production-to-shipment cycle. All these techniques of supply chain management are intended to continuously rebalance the multiple production and distribution activities.

This continuous rebalancing involves four elements:

- Strategy, which is concerned with the level of competitive response required from the chain in response to actual or anticipated challenges.
- Continuous refinement of the business process, which is concerned with adopting and implementing improved capabilities among all chain partners.
- Technology application, which is concerned with selecting and continuously renewing mutually compatible sets of enabling technologies simultaneously among chain partners.
- Human resource development, which is concerned with restructuring organizational relationships and retraining personnel involved in supporting the chain.

Outsourcing

Tradeoffs between insourcing and outsourcing essential functions within supply chains are continuously shifting, primarily in favor of outsourcing as transaction and search costs continue to decline. The Internet is driving these tradeoffs and enabling new forms of outsourcing to be realized profitably. The relationship between first and second parties, who are fully vested partners in supply chains, and third-party vendors has also shifted. Increasingly, not only are directed functions being outsourced, but so are entire processes whose successful performance requires high degrees of decisionmaking autonomy from third parties.

As more functions are outsourced to geographically dispersed agents, the virtual chain or the networked organization is becoming a reality. The strategic thinking that underlies these developments is to invest only in competencies that give a company a competitive advantage and to outsource to the top one or two third-party providers all other functions in which the company runs behind the level of cost, efficiency, or dynamism available from third parties. In this way a company ensures that it is creating, not losing, its competitive lead over other companies operating in its markets. The core functions typically retained by companies that are also integrators within their chains include planning, control, and recruitment and management of supply chain participants. Examples of Xerox, Dell Computer, Cisco Systems, Toyota, and Wal-Mart were cited above.

Globalization

Global market access is available today not only to extremely large corporations, but to small companies as well. Internet companies by definition compete in the global marketplace to which the Internet gives them direct and immediate access.

The primary challenge facing global corporations and constraining their worldwide market development is logistics. All parts of the globe are not equally accessible—or not accessible at an equalized price.

Transport service providers have carved up the globe into a set of common markets to which access is possible through regional distribution and production platforms. In most cases, transport price equalization within these markets reinforces their integration. Postal services were originally organized to establish common markets within national boundaries. The development of regional and global markets is increasingly left to global transport providers such as United Parcel Service, Maersk Moeller, Lufthansa, and DHL.

Transit time and cost rather than distance or national boundaries increasingly define common market parameters, and the parameters of competitive markets increasingly conform to the operational hub, spoke, or other network structures of the global or regional transport service providers serving these markets.

Improved Methods of Supply Chain Management

The competitive advantage of supply chains derives from a simple principle: collaboration among supply chain partners allows the best operational competencies and the best market access of individual members of the supply chain to be leveraged for the benefit of the entire chain. The compound leverage of combining individual organizations that are all "best in their class" is the primary source of competitive strength for the chain. However, the balancing of multiple interests and the unequal sharing of risks and rewards make many supply chains inherently unstable. Participants in chains are collaborators, but they may also be potential competitors.

As the science of industrial engineering continues to advance, more knowledge about the design and configuration of supply chains is being tested in applications and revealed to supply chain managers. In designing supply chains, industrial engineers attempt to balance product flow rates across multiple organizational frontiers to ensure that global "best outcomes" can be sustained. They design corresponding information flows and control feedback systems to create appropriate incentives so that local managers within the chain will have the incentive to strive for global best outcomes. Optimum global outcomes are not always easy to achieve because complex supply chain systems are unstable. Matching production and demand in competitive environments subject to multiple perturbations is tricky work. Substantial progress continues to be made, however. One area of notable achievement is forecasting of demand within the entire supply chain and analysis of production schedules that explicitly trade off the cost of excess inventory and the opportunity cost of lost sales.

The development of a common language and a common syntax for supply chain activities by the Council of Supply Chain Management and its adoption as a common standard by commercial software developers should greatly facilitate the flexible recruitment and engagement of supply chain partners. Additional commercial applications software for projecting demand and monitoring inventory at separate points in complex channels also extends opportunities for active supply chain management with interchangeable partners.

In addition, risk management techniques continue to evolve, and opportunities are continuously expanding to measure precisely defined risks that cannot be effectively managed within the chains and then sell them to risk arbitrageurs outside the chains. Technical advances in product design also continue to contribute to effective supply chain management. Increasingly, products are designed to minimize not only the cost of fabrication and delivery, but also the cost of reclamation and remanufacture. As noted above, Xerox Corporation—a large office equipment producer—designs most of its component subassemblies for reuse in second- and even third-generation machines.

Advances are also being made in organizational design that allow the necessary human response times and adaptability to be built into supply chains. Loose network structures that value flexibility and agility over hierarchical command and control are replacing more formal vertical or horizontally integrated structures and decentralized and horizontally affiliated structures in supply chains.

Supportive Public Policy

The most important factor enabling supply chains to develop within an economy is public policy. Supply chains are quasi-public goods. The benefits from the formation of supply chains redound to all participants in a particular trade or commercial sector. Supply chains also create positive externalities for competitors and participants in parallel supply chains. Moreover, the costs of developing supply chains typically exceed the benefits realized by any individual participant or any individual chain partnership. For this reason supply chains, like other public goods, are frequently underinvested. Thus it is appropriate for governments to facilitate the development of supply chains or to create institutions enabling most of the beneficiaries to capture the full measure of social benefits of supply chain development.

Another reason for public intervention is that many of the benefits and beneficiaries of investment in supply chain development reside beyond the borders of national economies. Thus the jurisdiction and corresponding incentive to invest in supply chains of individual countries may not subsume the full measure of potential supply chain benefits. Like markets, supply chains have some of the attributes of global public goods.

Few governments focus on supply chain development as a policy or investment priority. Most aspects of public policy that affect supply chain operation have accumulated incrementally through accretion. The effects of old policies or regulations are manifested most dramatically in the failure of supply chains to form or to take hold in a specific economy. Often a zero-based review is undertaken only on the occasion of a wholesale review of transport, trade, and trade process policy in the context of implementing a regional trade or common market treaty.

Moreover, investment in supply chain development is multifaceted. It includes various dimensions, some more amenable to private investment and to public sector investment and development. And some of these dimensions are more subject to government regulation and control than others. Figure 8.2 ranks the multiple aspects of investment in supply chains by their degree of difficulty to implement.

These same dimensions can be used to describe their responsiveness to government investment or other interventions. Supply chains typically develop in response to an explicit business strategy. Providing this strategy and implementing it are clearly a private sector responsibility. Government policy has minimum leverage on this aspect of supply chain development. Developing organizational structures in which to install supply chain functions is, however, a matter on which public policy can have a substantial effect. In many countries business tax law, business licensing procedures, company law, and commercial practice interact either to constrain or to enable various business models and to determine which are most suitable for supply chain functions.

At the next level of the pyramid, the selection and use of best-in-class production and distribution technology is primarily an area of private sector discretion unless government policy prohibits specific imports or differentiates duty levels by technology classes or sources of supply.

Supply chains require public infrastructure as conduits for transporting and transmitting goods, information, cash, and ownership rights. The conditions and user fees for using infrastructure are frequently determined by the public sector. Similarly, trade policy and trade services directly affect the efficiency with which supply chains can operate across national borders. Public policy substantially affects both dimensions.

Figure 8.2 Supply Chain Development Is Multifaceted

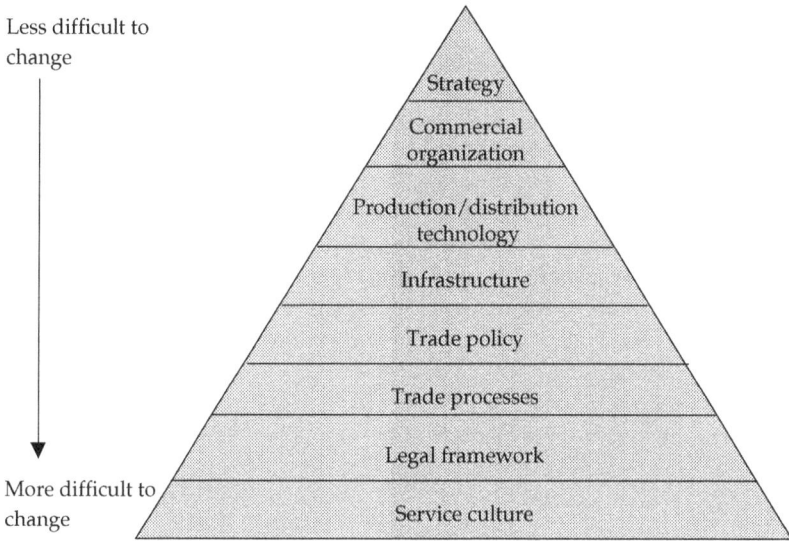

Less difficult to
change

Strategy

Commercial
organization

Production/distribution
technology

Infrastructure

Trade policy

Trade processes

Legal framework

More difficult to
change

Service culture

And it affects the legal framework in which contracts are interpreted and enforced, liabilities clarified and enforced, securities protected, insurance regulated, and prices equalized among classes of customers. Indeed, the primary role of government in fostering supply chains is to clarify the rules under which they must internalize transaction costs and that ensure that all supply chains compete on an equal basis.

The final and most ephemeral dimension involves the service culture in which supply chains operate. Service cultures are affected by many antecedent conditions including the level of civility and the tradition of commercial transparency within an economy. Government may have some effect on service culture but only indirectly and over the long term.

References

Christopher, Martin, and Lynette Ryals. 1999. "Supply Chain Strategy: Its Impact on Shareholder Value." *International Journal of Logistics Management* 10 (1):1–10.

Christopher, Martin, and D. Towill. 2000. "An Integrated Model for the Design of Agile Supply Chains." Paper prepared for the Logistics Research Network Conference, Cardiff, September 2000.

Shah, Janet, and Nitin Singh. 2001. " Benchmarking Internal Supply Chain Performance: Development of a Framework." *Journal of Supply Chain Management.*

Useful Websites

http://www.apics.org/scripts/bookstore/catalogfind.asp?id=1407&title=true

http://www.ewita.com/

http://www.supplychaintech.com/default.asp?section=sctnthismont

http://www.loggie.com/search.asp?query=Rail+Freight&ft=Rail&mode=m&cf=s

http://www.foodlogistics.com/foodlogistics/index.jsp

http://www.manufacturing.net/scm/index.asp?layout=siteInfoWebzine&view=Detail&doc_id=29997

http://www.driving-innovation.net/default.asp

http://www.ijlm.org/

http://www.napm.org/Pubs/journalscm/JournalArticleIndex.cfm#top

9

Protecting Intellectual Property: Why, How Much, How?

Manjula Luthria

A contentious aspect of globalization is the universalization of protection of intellectual property rights (IPRs), now required of all signatories to the Trade-Related Intellectual Property Rights (TRIPS) Agreement under the World Trade Organization (WTO). Under TRIPS all signatories must offer statutory protection to all technologies, product as well as process (Box 9.1).

There are several reasons for this unease over intellectual property rights. Developing countries view themselves as users rather than generators of intellectual property, so protection of intellectual property rights has not been a high priority for them. Reforming intellectual property rights regimes in poor countries is in fact seen as premature—much like adopting Western labor standards or environmental protection laws. Moreover, the TRIPS agreement represents the first time that domestic policies are being harmonized under the WTO, and developing countries are concerned about this expansion of the WTO's purview.[1] Finally, requiring poor countries to adopt stronger intellectual property rights appears to deny them instruments of economic development that have been successfully used by many industrial countries.[2]

Although some development experts have lobbied intensely for removal of intellectual property rights from the WTO, that seems unlikely to happen. So how should developing countries approach their policy options? Is there room for tailoring the new intellectual property rights regimes to the development needs of poorer nations? How should implementation proceed so that developing countries can not just come into legal compliance, but also fully exercise the leeways in implementation that exist? What complementary policies might they consider that could attenuate some of the costs while enhancing the potential benefits of stronger intellectual property rights? Developing countries will only exacerbate the "damage" if they view strong intellectual property rights as unfair and pro-rich, while neglecting to take advantage of the few but potent policy options that remain to allow customization of intellectual property reform to their development needs.

Box 9.1 Summary of Obligations under Trade-Related Intellectual Property Rights (TRIPS)

TRIPS sets out the minimum standards of protection to be provided by WTO members in each of the main areas of intellectual property:

- *Patents and the protection of new plant varieties.* Patent protection is provided for all products and processes in all fields of technology; minimum protection is 20 years from filing date. Domestic production can no longer be required, and restrictions are placed on compulsory licensing. Exceptions are allowed for plants and animals developed by traditional methods; an equivalent sui generis system protects rights of plant breeders.
- *Copyright and related rights* (for example, the rights of performers, producers of sound recordings, and broadcasting organizations). The minimum term is 50 years. Software programs are protected as literary works.
- *Undisclosed information* (trade secrets and test data). Trade secrets are protected against unfair methods of disclosure; test data submitted for regulatory approval of new pharmaceutical or agricultural chemical products are protected against "unfair commercial use."
- *Trademarks and geographical indications.* International agreements are further clarified; stronger protection is given to wines and spirits.
- *Industrial designs and integrated circuits.* Protection is for 10 years.

TRIPS also requires domestic procedures and remedies for enforcement of intellectual property rights. Disputes between WTO members over compliance with obligations are subject to the WTO's dispute settlement procedures. The obligations under the agreement apply equally to all member countries, but developing countries have a longer period to phase them in. For least-developed countries the compliance date is January 1, 2006.

Intellectual Property Rights 101, An Introduction

Patents are territorial rights: inventors must apply for a patent in each country in which they want protection from imitation. Pure ideas or concepts— scientific laws, formulas, theorems—cannot be patented. To be patentable, a product or process not only must be new and nonobvious, but also must possess industrial applicability (although the patent holder bears the risk of commercial viability). A country's patent office searches worldwide to satisfy itself that these criteria are met. Once preliminary searches have been completed, the patent application is typically published in an official gazette (this stage is called public disclosure). If no objections are received, the patent application is accepted and granted. The duration of patents varies considerably. A renewal fee is due at regular intervals to keep the patent alive. If the fee is not paid, the patent is voided. Most patents do not last for their full period of validity. In most countries, a majority of owners choose not to renew their patents beyond year five or so.

The patent application process is expensive and can take several years. All patent offices maintain detailed records of patented inventions. These are in the public domain and can be viewed by anyone. After a patent is granted, the burden of enforcement of the right is on the patent owner.

Why Protect?

In many ways the arguments for the protection of physical property apply to intellectual property as well: exclusive but limited rights of ownership need to be awarded to encourage investments in the creation and improvement of physical or intellectual property. Since knowledge possesses the additional characteristic of imperfect appropriability (unlike physical property, knowledge cannot be "owned" exclusively), investments in the creation of knowledge are believed to be curtailed without the help of instruments that can correct this inherent market failure. Thus copyrights protect the rights of authors (books, music, software),[3] trademark registration protects trade logos and symbols, and patents protect inventions with industrial applicability (products as well as processes). For technology development it is patents that seem to provoke most of the debate on technology-driven competitiveness. Therefore this section focuses on the unique issues surrounding this instrument.[4]

Patents are supposed to spur innovation in a number of ways. First, they award exclusivity of use, sale, and manufacture to the owners of the intellectual property, thus compensating them for undertaking expensive and risky innovative activities. In exchange for this benefit, the owner must disclose the invention on the patent document for anyone "skilled in the art" to be able to replicate. Thus patents are a tradeoff: a market distortion in the form of a temporary monopoly is created in exchange for disclosure of the information relating to the technology. Disclosure is intended to benefit society by disseminating new technologies and indeed encouraging competitors to invent around the new technology in a second round of innovation.

Advocating stronger intellectual property rights therefore presumes that the combined positive impact of the appropriability incentive for the innovator and the disclosure element for peers outweighs the negative impact of the temporary market distortion, thus making intellectual property rights protection beneficial to society as a whole. This is nearly impossible to test empirically and remains a highly debated subject.

How Much to Protect?

The evolution of intellectual property rights regimes has tended to exhibit a U-shaped relationship between income levels and intellectual property rights protection (Figure 9.1) Thus policymakers in developing countries had little reason to expect a need to strengthen their regimes so soon. Low-income countries (mostly former British and French colonies in Africa) possessed strong intellectual property rights regimes in the absence of a domestic imitative-industry lobby (they lie in the upper left side of the curve). High-income economies have protected intellectual property rights very strongly (they lie in upper right side), and middle-income countries offered the least protection for intellectual property rights (they lie at the bottom of curve) on the rationale that technological learning and catch-up are facilitated through minimal protection.

There is also a theoretical reason not to expect an across-the-board strengthening of intellectual property rights for most developing countries. Since intellectual property rights are meant to induce innovation by ensuring proprietariness for the results of research and development (R&D) while also imposing a welfare cost on society by creating a monopoly, protection should be thought of in terms of "optimal" rather than "maximum." Add to this a belief in diminishing returns to R&D, and the time will come when, at the margin, the global impetus to innovation is less than or equal to the

Figure 9.1 Intellectual Property Rights and Income

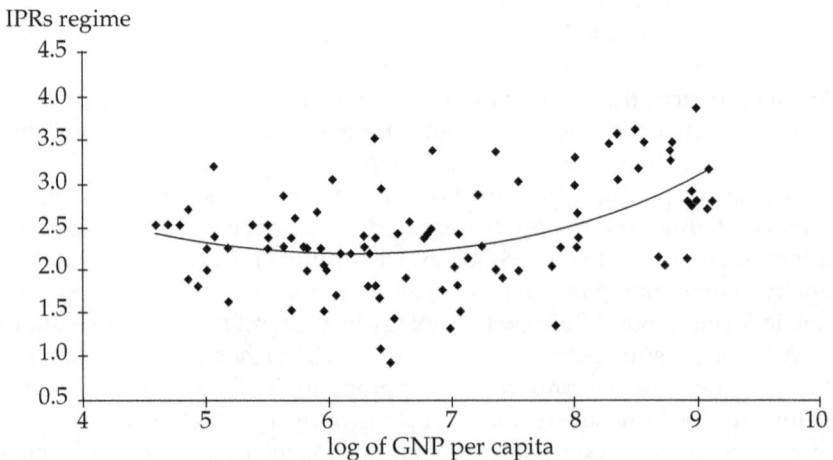

Note: IPRs regime is measured using Rapp and Rozck (1990) Index.

loss in consumer welfare associated with it.[5] That implies that not all countries would need to join the strong intellectual property rights club in order to induce greater global innovation (Deardorff 1992).

However, two important events are changing the shape of things to come. First, investment flows are seeking destinations throughout the world, and the ability of multinational firms to protect their knowledge assets is becoming a critical determinant in choosing a destination. A regime that protects the appropriability of the returns to the knowledge assets of a multinational firm is believed to make the difference between whether a firm is able to position itself as a leader in new markets or loses out to the lowest-cost imitator in the host country. Second, as pointed out above, all WTO members who are signatories to the TRIPS agreement have to upgrade their intellectual property rights regimes.

This is a new era for all parties involved. Industrial countries are eager to install instruments to protect the knowledge assets of their domestic firms, expecting significant gains to their investments. Developing nations are also becoming increasingly sensitive to the needs of the new knowledge economy and eager to reap the benefits of stronger intellectual protection but are unsure about how to proceed.

What Are the Claims—and Fears ?

Both benefits and costs are identified in the debate over stronger intellectual property rights.

Benefits

One of the principal claims of benefits from universally strong intellectual property rights is an increase in inventive activity. By analogy with physical property rights, it is logical to assume that greater appropriability of returns from inventive activity will induce greater inventive activity. However, evidence on the effect of a strong patent regime on R&D investments is mixed. Firms rely on a host of instruments for appropriating the returns to their risky and expensive activities, and patent protection has not tended to determine their level of effort.[6] Of course, such reliance varies greatly by sector. Patent use tends to be greater in sectors such as chemicals, where imitation is relatively easy, than in sectors such as automobiles, where natural barriers to entry are high.

The claim that stronger intellectual property rights could result in greater technology transfer to developing countries may carry more weight. Greater contractual certainty about proprietary knowledge is likely to facilitate the contracting out of technologies through licensing or foreign direct investment (FDI). Both licensing and FDI are sensitive to an environment that protects the knowledge assets of the owner, so a secure environment is likely to be more conducive to making such transfers.[7]

Another claimed benefit is of greater trade for countries with stronger intellectual property rights. In reality, the effect is ambiguous. Imports could increase as foreign intellectual property owners face rising net demand for their products because "pirates" have been displaced, but a patent holder might instead reduce sales in a foreign market because it has greater market power at home in an imitation-safe environment. Whether market expansion or market power effects will be stronger is not known.

Finally, there is also the expectation, regardless of whether aggregate R&D increases or not, that R&D in industrial countries is more likely to incorporate the needs of developing countries (malaria and typhoid control, higher yielding rice varieties) if researchers could be guaranteed a return on their investments. More than intellectual property rights protection is likely to be needed. For example, drugs would need to be priced quite high to earn profits, so in addition to proprietariness, pharmaceutical firms would need to be assured that sick consumers in developing countries could afford the cures developed in industrialized countries. By itself, then, stronger intellectual property rights protection is unlikely to be the miracle cure for endemic developing world problems.

Costs

While claims for the benefits of stronger intellectual property rights rely on strong assumptions that cannot be tested immediately, the costs that stronger rights impose on developing countries are likely to be real and quite immediate. These immediate costs and delayed benefits are the main reason for the difficulty in building political will and national consensus on stronger intellectual property rights.

The most observable immediate cost is likely to fall on consumers through increased prices of patented technologies—mainly in the pharmaceutical sector. Since patents matter a great deal to this sector, manufacturers are likely to protect every patentable invention. Most patents taken out by multinational corporations in developing countries are in this sector, and the high price of essential pharmaceutical products is a grave concern in most poor nations.[8]

In response, pharmaceutical manufacturers have argued that the pricing of their products in poor nations is likely to take into account the low incomes and weak insurance markets. This remains to be seen, but it can hardly be ignored that drug prices are lowest in countries where patent protection is weak (India, for example; see Watal 2000b).[9] And because drug prices in industrial markets are regulated using global reference pricing, patent-owning firms may choose to sell in developing country markets at substantially higher prices than would maximize local profits because they do not want to jeopardize the prices at which they are allowed to sell in other regulated markets.[10]

Another immediate cost is likely to be the effect on local industries that thrive on imitation. It is argued that pirate or imitative industries that are displaced after the adoption of intellectual property rights could turn to producing off-patent products (products or processes using patents whose legal rights have been allowed to lapse by patent owners or whose full term has expired). The off-patent market is much larger than the patent market in most developing countries, implying a huge untapped market segment for local industry. In reality, however, a firm that has secured a first-mover advantage in introducing a patented product enjoys a significant lead in marketing its own generic version once the patent expires.[11]

Important concerns also surround the awarding of title rights for agricultural research, which threatens to shift research from the public to the private sector, thus raising concerns about the diffusion of agricultural technologies and possibly threatening food security in poor countries. The new focus on miracle technologies and copyrighted, branded solutions to food problems linked to proprietary gains is being criticized as too narrow an approach to agricultural development. There are also fears that patents will induce the commercialization of farming along the lines of farming systems in industrial countries, thereby undermining small-scale subsistence and local market-based production systems.

Food security for most small farmers rests on their access to land, water, seeds, and tools, making complete dependence on the market for their inputs too risky. Also, some seeds and plants may not be amenable to intellectual property protection, thus distorting the availability of certain agricultural products. Much-publicized attempts to patent age-old knowledge that has been held tacitly and passed down for generations (such as Indian and Chinese herbal medicines and Kente weave designs from Ghana) have generated great mistrust. Simple indigenous agronomic practices (interculturing, distance between rows, conventional breeding, biopesticides) are being threatened with removal from the public domain to proprietary territory. Small firms may find it increasingly difficult to enter new businesses in the face of the assembled patent rights of industry leaders and may find contract restrictions on access to marketed materials that would once have been freely available for further breeding (Tansey 1999).

Finally, compliance with TRIPS entails direct administrative costs. Patent offices are expensive to run. While developing countries need not duplicate Western models, substantial resources will have to be devoted to administration of these offices. Related costs (building awareness and training judges, lawyers, and enforcement personnel) and indirect costs (skilled labor pulled from other activities) are also expected to be burdensome for many poor countries.

Five Ways to Swallow Bitter Medicine

Stronger standards for protection of knowledge assets are now an integral part of globalization. For understandable reasons on both sides, the dialogue in recent years has been clouded by emotional arguments, with scant attention to ways to customize the new standards to suit specific country requirements or to the use of policy levers to complement these instruments in promoting innovation. There is an urgent need to guide developing countries in building up their innovation systems, taking full account of their national developmental goals and international obligations in the area of intellectual property and trade. This section summarizes the levels of engagement that government and industry could choose to follow in doing their part to mitigate the burden of stronger intellectual property rights while buttressing some of the benefits. It presents five ways to approach intellectual property rights reform.[12]

Because the Doctor Says So

One approach is to assume that stronger intellectual property rights are inevitable and to go ahead and adopt the stronger standards. It would not be an exaggeration to say that this attitude characterizes the sentiment in many developing countries. Most view it as a battle that developing countries could not win in the international arena—particularly against the powerful industry lobbies of the West. Worse, most developing countries describe their state of preparedness on the subject as weak at the time of negotiations. Also, there is a sense that stronger intellectual property rights are the "price" of improved access for textiles. Predictably, this view has led to little progress on the how-tos of implementation.

It'll Do You Some Good

Another approach starts with a belief (with necessary qualifications) in some of the broad claims made for stronger intellectual property rights. For example, evidence shows that low-technology trade is sensitive to the intellectual property rights regime, much more so than trade in medium- or high-technology goods (Braga and Fink 1999). Medium-technology products tend to possess high barriers to entry, which prevent easy imitation (automobiles are a typical example), and high-technology products (electronics) have considerably shorter product cycles for which first-mover advantage matters most (for this reason some firms even substitute FDI for trade). Thus, for a large majority of developing countries that are just entering technology-based trade, intellectual property rights may be more relevant than they might think (Table 9.1).

For foreign investment, the intellectual property regime would be important not only for attracting FDI to knowledge-sensitive sectors (Mansfield 1995), but also for obtaining the spillovers from FDI in the form of technology improvements for domestic industry at large. To facilitate spillovers from FDI in the form of tacit transfers of new technologies, human capital, or superior management techniques, local firms need to be engaged with multinational firms in a subcontracting relationship (Battat, Frank, and Shen 1996). Subcontracting of production processes with proprietary elements is most likely to be facilitated by contract certainty. The same argument is valid for deepening the supply chain—whose creation and management are becoming a source of significant efficiency gains for firms (see Chapter 8).

Follow Bitter Medicine with Sweeteners

Some aspects of the patenting process can be tweaked to align patents more closely to technology diffusion as opposed to technology-generation goals. Japan, during its early development stages, carefully designed its patent system to suit the needs of a technology user rather than a technology generator. Thus, for example, systems can call for public disclosure of an invention earlier than is customary in many industrial countries. Countries that fear that "everything will be patented" can pay more careful attention to the

Table 9.1 Regional Shares of Developing Countries' Manufactured Exports by Technology Intensity

(Percent of developing world exports)

Intensity of technology	Year	East Asia	South Asia	Middle East and North Africa	Latin America	Sub-Saharan Africa
Resource-based	1985	34.6	3.8	23.8	32.9	—
	1998	47.5	4.7	15.0	28.0	4.8
Low technology	1985	71.7	8.3	7.3	11.9	—
	1998	70.2	8.5	7.2	12.6	1.5
Medium technology	1985	63.4	2.0	7.1	25.8	—
	1998	63.8	1.8	4.4	28.1	1.9
High technology	1985	81.0	1.1	1.8	14.8	—
	1998	85.5	0.6	0.7	12.9	0.4

— Not available.
Source: Lall, 2000.

criteria for granting patents. Countries need not follow the U.S. model of granting patents readily and waiting for dissent to correct any problems, although that would increase the burden on examiners.[13] Another approach is to limit the scope of patents, thereby reducing the market distortion caused by awarding proprietary rights to new technologies.[14] Differential pricing could be used to support the interests of small firms or individuals filing for patents.[15] Attention to the design of patent offices, infrastructure, and personnel training to make access to patents on a worldwide basis less cumbersome would also encourage dissemination.

More substantially, countries could add instruments such as utility models that are well suited for newcomers on the innovation scene. Utility models, or petty patents as they are often called, have substantially lower criteria for demonstrating novelty in exchange for shorter periods of protection—usually five to seven years. Typically, small improvements and modifications to existing technologies qualify for protection, which makes utility models particularly well suited for encouraging local innovation.[16] While "absolute novelty" is necessary for filing for a patent, a lower level of inventiveness is required for a utility model. And registration is quicker and simpler without the need for examination of novelty and level of inventiveness—on average six months instead of up to four years for a regular patent. Quick registration also enables more rapid commercial exploitation of the invention, whether through licensing or direct use. Smaller firms are the main beneficiaries of such protection since they engage mostly in adaptive innovation.

There are also a few policy options for addressing concerns about higher prices as a result of the monopolies created in protecting intellectual property rights. To begin with, fear of monopoly pricing is exaggerated since markets in developing countries are not currently characterized by competitive pricing. Therefore, encouraging competition by streamlining entry and exit policies and adopting competition law policy is the first place to start. Competition law can deal effectively with the regulation of monopoly power and the containment of abuse of dominant position (Box 9.2 and Chapter 3). Competition policies can also deal fairly with issues of compulsory licensing for sensitive technologies. Price regulation, where appropriate (in health matters, for example), is another policy option, but it would need to be carefully considered because of its potentially adverse effect on the incentive to encourage competition.

Pursue a Holistic Approach

Ultimately, strong intellectual property rights will mean as much or as little to developing nations as their capacity to engage in innovative activity. Innovative capacity is the outcome of several factors—the internal deci-

Box 9.2 Competition Policy and Intellectual Property Rights

At the broadest level intellectual property rights and competition policy are complementary because they share a concern to promote technical progress to the ultimate benefit of consumers. Firms are more likely to innovate if they are protected from "free-riding." They are also more likely to innovate if they face strong competition. Yet there is often a perception that the two policies are in conflict. This perceived conflict stems from three areas of uncertainty: the extent to which competition policy is about short-run allocative efficiency or long-run dynamic efficiency, whether market power should be inferred from the existence of an intellectual property right, and whether a particular contract, license, or merger should be regarded as horizontal or vertical.

In some areas, however, competition authorities have been clear about taking action against intellectual property right holders who try to use intellectual property rights to:

- Coordinate a cartel, such as through patent pooling of substitute technologies.
- Create economic advantages outside of the market where the innovation took place.
- Prohibit use of licensed technology after the license has terminated or require royalty payments for a term exceeding the life of the patent.
- Prohibit a licensee from challenging the validity of the patent.
- Refuse to license or insist on excessively high royalties that inordinately restrict competition, such as when the intellectual property holder owns an "essential facility."
- Patent a wide variety of competing products and processes that are never intended for use or sale ("killer patent portfolios").
- Establish mergers that create monopolies not just in product markets, but also in the innovation market.
- Institute bad faith litigation to prey on competitors, particularly small firms.

Finally, an area that is likely to cause considerable debate at the interface of intellectual property and competition policy is the treatment of parallel imports (goods brought into a country without the authorization of the intellectual property holder after the goods were placed legitimately into circulation elsewhere). Policies regulating parallel imports stem from the territorial exhaustion of intellectual property rights. Under national exhaustion, rights end on first sale within a nation, but owners of intellectual property rights may prevent parallel trade with other countries. Under international exhaustion, rights are exhausted upon first sale anywhere, and parallel imports are permitted.

sions of firms, the existence of a support infrastructure, and the right mix of skills.

While the firm must make the decision to engage in improving technologies, public policies can help create an environment that is conducive to making such investments a priority (Chapter 5). All governments have instituted some incentives for encouraging such efforts, but their deployment and effectiveness vary considerably. An environment conducive to innovation requires an array of research and technology institutions that can supply such basic services as quality certification and testing or provide more upstream services such as contracting out specific R&D functions. Most research institutions require some form of public support, although encouraging self-reliance seems to be important to the effectiveness of their services. Finally, efforts to build and train human capital through the traditional education system and through enterprise commitment to the training of workers are also prerequisites to creating an innovation-encouraging society (Chapter 6). Without these ingredients, strengthening intellectual property rights is unlikely to yield any immediate positive benefits by way of technology generation or transfer.

Be Your Own Doctor

Many intellectual property rights issues are still incompletely understood. Identifying, protecting, and if necessary commercializing indigenous knowledge are new areas for many societies. Each country may need to find its own "green gold" and market this know-how in a manner that remunerates local people. New instruments may need to be explored for this purpose, such as the local innovation registries that some farming communities have adopted.

The human genome project has brought to the fore concerns about ownership of genes and the debate over the proper instrument—copyrights or patents—for protecting software and business processes like the Amazon one-click, to name just a few, challenge our understanding of how intellectual property rights should be awarded. Developing countries feel marginalized in discussions of such issues, in part because of a sense that these are state-of-the-art or Western problems. But the repercussions are worldwide, and developing countries might better serve their interests by becoming engaged in the dialogue.[17]

Concluding Remarks

International development institutions seek to promote economic growth in poor countries, ensuring that the poor participate in the growth process as well as providing direct assistance to the most vulnerable sections of society. Intellectual property rights have a bearing on each of these goals. Not only do intellectual property rights directly affect economic growth through their impact on innovation or imitation-led catch-up, but recent developments could potentially threaten access to proprietary knowledge for the poor, thus raising new equity and participation concerns. Because reduced access to world-class technologies for developing country populations could impede the mission of poverty alleviation, international development institutions need to be pay greater attention to intellectual property rights than they have thus far.

A credible nonpolitical entity is needed to help provide an impartial methodology for solving some of the difficult issues outlined above. As a start, considerable analytical work is needed to fill the knowledge gaps on the interplay among intellectual property rights, markets, competition, and investment attraction in developing countries so that a coherent set of benefits and costs can be identified. Case studies of the performance of East Asian economies that have fared well in terms of innovation and FDI-attraction before their intellectual property rights regimes were formally upgraded could answer some questions and shed light on the institutional underpinnings of innovation. This analysis could lead to the development of complementary mechanisms for fostering innovation-led growth.

Core constituencies in developing countries also need to be apprised of impending changes so that private industry, government, consumers, agriculture, and the legal community can become engaged in the dialogue and understand the ramifications of the new regime. Identifying key obligations and raising awareness among key constituencies are crucial for ensuring that intellectual property rights reform is consistent with a country's stage of development (to the extent compatible with WTO commitments). Such extensive involvement will ensure fuller and more effective participation from developing countries in future rounds of negotiations on related matters.

Once the key concerns of the most-affected sectors are fully understood and documented, customized services and advice on TRIPS-compliant policies will need to be developed on a country-by-country basis as implementation deadlines near. Policy instruments that reduce the tension between equity considerations and proprietary motives—privatization, competition policy, price regulation, licensing, targeted subsidies, transfer and redistribution mechanisms—will need to be devised, and best practice in design and enforcement disseminated. International development institutions are well positioned to broker this dialogue between various factions and to help dissipate some of the confusion that surrounds the protection of intellectual property.

Notes

1. Intellectual property rights are territorial in nature and have been awarded and enforced within the confines of the domestic policies of a nation.

2. For instance, many industrial countries excluded pharmaceutical products from protection until recently: Switzerland until 1977, Italy until 1978, Spain and Norway until 1992, Finland until 1995, and Iceland until 1997. Liberal compulsory licensing policies have been followed in Canada, France, and the United Kingdom.

3. Note that copyrights only protect the expression of the idea or concept from plagiarism, not the underlying idea itself from being applied. Patents protect the underlying idea or invention from being copied. Copyrights are therefore not seen as hampering the application or diffusion of ideas contained in books, software, or music. This is the rationale behind copyrights lasting a very long time (50 or more years after an author's death) and patents only around 20 years from filing.

4. Copyrights for protecting the music and software industries or trademarks for signaling product quality and building consumer loyalty also affect competitiveness. For instance, Chinese producers reported difficulties in promoting their own brands of soft drinks, processed foods, and clothing. Having established brand recognition through costly investments in marketing and distribution, enterprises find their trademarks quickly applied to counterfeit products, damaging the reputation of high-quality producers and sometimes forcing producers to abandon their trademarks or even close down altogether (Maskus, Dougherty, and Mertha 1998).

5. The additional boost to global R&D obtained from stronger intellectual property right protection in, say, Cambodia, is likely to be negligible compared with the costs imposed on Cambodian consumers.

6. The appropriability ratio of R&D investments through patents (ex post subsidy to "successful" R&D) is estimated to be in the range of 15–20 percent in industrial countries (Lanjouw 1993), meaning that 15–20 percent of industry's R&D investments are recouped through patents in industrial countries. Comparative estimates for a developing country exist only for India (Luthria 1996), where this ratio is less than 2 percent.

7. In surveys German, Japanese, and U.S. firms report that the strength or weakness of an intellectual property rights regime has an important effect on some, but not all, types of foreign direct investment (Mansfield 1995). About 80 percent of firms maintained that it was important to investment in R&D facilities, whereas about 20 percent thought it was important for investments in sales and distribution facilities.

8. Forecasts of price rises depend on the type of market structure that exists prepatent and the elasticity of the demand curve. Simulations have estimated a price rise of from 20–205 percent (Watal 2000b). However, to capture any price dif-

ferences due to patent protection accurately, a price index of a basket of new and unpatented drugs has to be compared with a basket of new patented drugs. Such an experiment, which has not yet been conducted, would need to control for the relative quality of the new and old drugs as well as their therapeutic substitutes. Some evidence shows, however, that average product prices fall sharply when generic entry occurs following expiration of patents. Caves, Whinston, and Hurwitz (1991) showed that in the United States the average price of generic substitutes was 60 percent that of the branded drug with 1 entrant, 29 percent with 10 entrants, and 17 percent with 20 entrants.

9. It should be noted that low pharmaceutical prices in the Indian market result not only from weak patent protection, but also from the intense competition that characterizes this sector. Of course, it could be argued that competition is itself a consequence of a past weak intellectual property right regime, but the point is that in the absence of a vibrant domestic industry, simply not having patents will not result in such low prices.

10. For countries that fix ceiling prices, the price of a new drug is related to its price elsewhere. The 1993 Health Security Act in the United States proposed using the lowest price in 22 other countries as a criterion for determining fairness in pricing (Lanjouw 1997).

11. In fact, some observers have commented that it is the off-patent market in developing countries that U.S. firms want to seize rather than the on-patent market, particularly since competition from generic drugs in the U.S. market is steadily eroding profits.

12. The author acknowledges inputs from Keith Maskus of the University of Colorado and Jayashree Watal of the Institute for International Economics.

13. The United States probably has the least stringent criteria for determining patentability. The United States grants 95 percent of patent applications, whereas Germany, for example, grants only 40 percent.

14. If a substance had several different industrial applications, a regime that permitted a very broad scope of patents would allow one patent to be filed collectively. Broad scope typically imposes more stringent limits on alternative or even complementary products and processes.

15. Under the TRIPS agreement, all procedures must follow national treatment, which implies that discriminatory pricing cannot exist between domestic and foreign applicants.

16. For instance, utility models in the Philippines encouraged successful adaptive invention of rice threshers (Mikkelsen 1984).

17. In addition, the text of TRIPS is scattered with "may" provisions that are clearly optional. There are several undefined, ambiguous terms such as "inventions," "new," "inventive step," "micro-organisms," "essentially biological," "effective," "unreasonably," "legitimate," to mention just a few, that must be interpreted at the national level (Watal 2000a).

References

Battat, Joseph, Isiah Frank, and Xiaofang Shen. 1996. "Suppliers to Multinationals: Linkage Programs to Strengthen Local Companies in Developing Countries." Foreign Investment Advisory Services Occasional Paper 6. World Bank, Washington, D.C. Processed.

Braga, Carlos A., and Carsten Fink. 1999. "How Stronger Protection of Intellectual Property Rights Affects International Trade Flows." Policy Research Working Paper 2051. World Bank, Science and Technology Thematic Group and the Energy, Mining, and Telecommunications Department, Washington, D.C.

Caves, Richard E., Michael Whinston, and Mark Hurwitz. 1991. "Patent Expiration, Entry, Competition in the U.S. Pharmaceutical Industry." *Brookings Papers on Economic Activity*, Special Issue. Washington, D.C.: Brookings Institution.

Deardorff, Alan V. 1992. "Welfare Effects of Global Patent Protection." *Economica* 59: 35–51.

Lall, Sanjaya. 2000. "The Technological Structure and Performance of Developing Country Manufactured Exports." QEH Working Paper QEHWPS44. Oxford University, Queen Elizabeth House, Oxford, U.K.

Lanjouw, Jean O. 1993. Patent Protection: Of What Value and for How Long? NBER Working Paper 4475. Cambridge, Mass.: National Bureau of Economic Research.

———. 1997. The Introduction of Pharmaceutical Product Patents in India: Heartless Exploitation of the Poor and Suffering? NBER Working Paper 6366. Cambridge, Mass.: National Bureau of Economic Research.

Luthria, Manjula M. 1996. "Intellectual Property Rights in a Developing Country: How Much and for Whom." Ph.D. dissertation, Georgetown University, Economics Department, Washington, D.C. Processed.

Mansfield, Edwin. 1995. "Intellectual Property Protection, Foreign Direct Investment, and Technology Transfer." IFC Discussion Paper 27. International Finance Corporation, Washington, D.C.

Maskus, Keith E., and Mohan Pennbarti. 1995. "How Trade-Related and Intellectual Property Rights in Developing Countries?" *The World Economy* 23: 595–611.

Maskus, Keith E., Sean M. Dougherty, and Andrew Mertha. 1998. "Intellectual Property Rights and Economic Development in China." Paper presented at the Southwest China Regional Conference on Intellectual Property Rights and Economic Development, Chongqing. Processed.

Mikkelsen, K. W. 1984. "Inventive Activity in Philippines Industry." Ph.D. dissertation, Yale University, Economics Department, New Haven, Conn. Processed.

Rapp, Richard T. and Richard P. Rozck. 1991. Benefits and Costs of Intellectual Property in Developing Countries. *Journal of World Trade* 24:75–102.

Tansey, Geoffrey. 1999. "Trade, Intellectual Property, Food, and Biodiversity." Quaker Peace & Service. *http://www.quaker.org/quno.*

Watal, Jayashree. 2000a. "Intellectual Property Rights in the World Trade Organization, and Developing Countries." London: Kluwer Law International.

———. 2000b. "Pharmaceutical Patents: Options for India." *The World Economy* 23 (5): 733–52.

Author Biographies

Ijaz Nabi led the Competitiveness Thematic Group in the East Asia Region of the World Bank and was Lead Economist for South Korea, Lao PDR, Malaysia, Myanmar, and Thailand. He managed the Bank team that prepared the adjustment lending operations for Thailand during the 1997 financial crisis. He is currently Sector Manager, Economic Policy, in the South Asia Region of the World Bank, a position he has held since May 2002.

Manjula Luthria coordinates the programs of the Competitiveness Thematic Group in the East Asia Region of the World Bank. She has worked extensively on research and operational projects relating to innovation and diffusion issues, with a special focus on the economics of intellectual property rights.

Sanjaya Lall is Professor of Development Economics and a Fellow of Green College at Oxford University. His publications include 30 books and monographs and over 200 articles on international investment, technology transfer and development, trade, and industrial strategy.

Behdad Nowroozi is the coordinator for corporate governance in the East Asia Region of the World Bank. He has worked on corporate governance reform in Korea and Thailand since the 1997 financial crisis in that region.

R. Shyam Khemani is Competition Policy Advisor in the World Bank's Private Sector Advisory Services Department. Prior to this he was the manager for the Business Environment Group in that department. He has also served as Chief Economist and Director for Economics and International Affairs of the Canadian Competition Bureau.

Geeta Batra is a senior private sector development specialist at the World Bank. She has worked extensively in the areas of enterprise training and firm productivity and training policies in developing countries, and has published several papers in those areas.

Hong Tan is a lead economist at the World Bank Institute, specializing in education and training reform and its effect on firm-level productivity.

Prasad Gopalan is a senior investment officer with the International Finance Corporation. He has been actively involved in financing private sector transactions in East Asia and the Pacific since 1998. He is also the regional coordinator for IFC's technology activities in East Asia.

Ron Kopicki is a supply chain advisor at the World Bank. His focus on supply chain development issues has evolved from his prior work on private sector competitiveness enhancement and on quality transport service delivery.

Index

NOTE: *b* with page number refers to boxes, *f* refers to figures, *n* refers to notes, and *t* refers to tables.

www.ingramcontent.com/pod-product-compliance
Lightning Source LLC
Chambersburg PA
CBHW031809190326
41518CB00006B/264

* 9 780821 351543 *